FINANCING AND CHARGES FOR WASTEWATER SYSTEMS

WEF Manual of Practice No. 27

Fifth Edition

Edited by Rocky Craley

2025

Water Environment Federation
601 Wythe Street
Alexandria, VA 22314–1994 USA
https://www.wef.org

ISBN 978-1-57278-479-6

About WEF

The Water Environment Federation (WEF) is a not-for-profit technical and educational organization of 30,000 individual members and 75 affiliated member associations representing water quality professionals around the world. Since 1928, WEF and its members have protected public health and the environment. As a global water sector leader, our mission is to connect water professionals; enrich the expertise of water professionals; increase the awareness of the impact and value of water; and provide a platform for water sector innovation. To learn more, visit www.wef.org.

This edition edited by Rocky Craley

With contributions by

Edward Barca, CPA
Thierry A. Boveri, CGFM
Thomas Brightbill, PE
Alex Bugbee
Collin Drat
Matthew Freiberg
Matthew Gregg, PE
Andrew J. Hall, PE
Jennifer Ivey, PE
Rupa Jha
Theresa Jurotich, PE
Will Kerr, CPA
Prabha Kumar
Edward Kwong
Henrietta Locklear
Brian Merritt
Philip Sapone
David Sayers
Nikolai J. Sklaroff
Jennifer Tavantzis
Matt Wittern, APR, PMP
John Wright, CPA
Vyomini Upadhyay, PE
Bill Zieburtz

Review provided by

Raj Chavan, PhD, PE, PMP
S. Rao Chitikela, PhD, PE, PEng, BCEE, ASCE Fellow
William (Bill) H. Davis, PE
Vedansh Gupta, EIT
Varenya Mehta, PE
Mahith Nadella, PE
Keval Satra, PE
William G. Stannard, PE, Chair Emeritus and Senior Principal
Jeff Tucker, CPFO
Alan Tucker, MA, PMP
Harry Zhang, PhD, PE, EWRI Fellow

Text originally prepared by the **Financing and Charges for Wastewater Systems** Task Force of the Water Environment Federation

Jon Davis, *Chair*

Andrea Boehling
Thierry A. Boveri, CGFM
Ann Bui
Eric Callocchia
Rocky Craley
Jeff Dykstra
Jennifer Fitts
Ryan Graf
Christine Anderson Gray, ENV SP
Matthew Gregg, PE
Matt Gsellmeier
Loveenia Gulati
Jack Habzda
Michael Hines, PE
Jennifer Ivey, PE
Brian W. Jewett
Theresa Jurotich, PE
Prabha Kumar
Saumya Kurup-Green
Ting Lu, PhD, PE
Bryan A. Mantz, CMC, CGFM
John Mastracchio
Hannah McLean
Robert J. Ori
Sudhir Pardiwala, PE
Brian Porter
Justin Rasor
Joseph Ridge
Eric Rothstein, CPA
Philip Sapone
Brian Shell, PE
Nicholas T. Smith, CGFM
Marc Walch
Anna White
John Wright
William B. Zieburtz Jr.

Under the Direction of the **Utility Management Subcommittee of the Technical Practice Committee**

Water Environment Federation
601 Wythe Street
Alexandria, VA 22314-1994 USA
http://www.wef.org

Manuals of Practice of the Water Environment Federation

The WEF Technical Practice and Publications Advisory Panel (formerly the Technical Practice Committee and the Committee on Sewage and Industrial Wastes Practice of the Federation of Sewage and Industrial Wastes Associations) was created by the Federation Board of Control on October 11, 1941. The primary function of the Committee is to originate and produce, through appropriate subcommittees, publications dealing with technical aspects of the broad interests of the Federation. These publications are intended to provide background information through a review of technical practices and detailed procedures that research and experience have shown to be functional and practical.

Contents

List of Figures

List of Tables

Preface

This 5th edition of *Financing and Charges for Wastewater Systems* provides an overview of current industry practices that should be considered for financing and establishing rates and charges for wastewater collection and treatment systems. In addition, this Manual of Practice reflects and is responsive to changes in the industry over the past 6 years that have occasioned a heightened awareness about wastewater utility financial management, the rapid rise in cost of operations and capital investment, and the increased scrutiny of equitable distribution of cost responsibilities across customer groups or classes. These events and changes, including tightening environmental regulation and replacement of aging infrastructure, have resulted in a pronounced increase in the cost of wastewater services. Though there have been several federal initiatives over the past 6 years to infuse funding to the wastewater sector, costs have outpaced available funding and assistance and challenge the affordability of this vital service. In this context, guidance on wastewater rate-setting practices and, particularly, the distribution of costs across user groups are of critical importance.

As individual wastewater utilities address their financial-management challenges, it is important that their practices for setting rates and charges be responsive to the unique circumstances and values of the communities they serve. Accordingly, this manual offers industry-accepted guidance in addressing these challenges without prescribing specific methods. In many instances, acceptable methodological alternatives are presented.

Financing and Charges for Wastewater Systems is structured to provide the reader with an overview of financing mechanisms for wastewater systems and a framework for development of wastewater rates and charges.

Authors' and reviewers' efforts were supported by the following organizations:

AECOM
Aqua-Aerobic Systems, Inc.
Ardurra
Black & Veatch Corporation
Carollo Engineers, Inc.
Central Contra Costa Sanitary District
Civitas Engineering Group
HR Green, Inc.
Los Angeles County Sanitation Districts

Pittsburgh Water
Raftelis
RC-WEE Solutions, LLC
Stantec Consulting
Sacramento Area Sewer District
San Francisco Public Utilities Commission
Vallejo Flood and Wastewater District
Water Research Foundation

1

Introduction

1.0 INTRODUCTION

This manual of practice (MOP) replaces and expands upon previous guidance on wastewater utility financing and rate-making in the 4th edition of *Financing and Charges for Wastewater Systems*. It reflects changes in the industry since 2018 that have heightened awareness about wastewater utility financial management, the rapid rise in the cost of operations and capital investment, and the increased scrutiny of equitable distribution of cost responsibilities across customer groups or classes. There have been many national and international events, the greatest of which was the global pandemic of COVID-19, that wreaked havoc on all lives, industries, and countries. While officially, the pandemic is over, its effects are still being felt in a number of ways in the wastewater industry. The pandemic was followed by years of supply chain issues and high inflation, affecting all sectors and types of costs, that still continue as of the writing of this edition. Additionally, there has been significant social and economic debate surrounding environmental justice and support to ensure that utility operations and infrastructure investment are fairly and equitably implemented across service areas. These events and changes, including more stringent environmental regulations and continued efforts to replace aging infrastructure, have resulted in a pronounced increase in the cost of wastewater services.

United States government initiatives, such as the Infrastructure Investment and Jobs Act, also known as the Bipartisan Infrastructure Law and the Low Income Household Water Assistance Program were implemented to assist utilities and customers with available funding and support. Other federal and industry initiatives, such as the Water Infrastructure and Improvement Act and the Water Environment Federation's Circular Water Economy promote the interconnectivity of water, wastewater, and stormwater services and stress the cost and infrastructure efficiencies gained by working together.

Still, within the past 6 years, costs have far outpaced available funding and assistance. The wastewater industry still faces a well-documented infrastructure funding gap that will continue to drive wastewater rate increases, challenging the affordability of this vital service (Economic Development Research Group for American Society of Civil Engineers, 2016). At the same time, many utilities delayed increases to rates and charges to alleviate further customer financial burdens during recent turbulent times, which only further exacerbated the funding issue.

In this context, guidance on wastewater rate-setting practices, particularly the distribution of costs across user groups, is critical. As wastewater utilities address their financial management challenges, their practices for setting rates and charges must be responsive to the unique circumstances and values of the communities they serve. Accordingly, this MOP offers industry-accepted guidance to address these challenges without prescribing specific methods. In many instances, acceptable methodological alternatives are presented. Similarly, the numerical calculations are intended to illustrate the presented methodologies and, as such, specific numerical values should not be used outside of the context of the example calculations presented.

Methodologies and examples presented in Chapters 5 through 8 guide the development of rates based on a cost-of-service analysis. The term *cost-of-service* does not simply imply that the rates are adequate to recover system revenue requirements (e.g., utility costs); rather, cost-of-service in the context of this manual refers to rates that have been developed using industry-accepted cost-allocation approaches to price utility service in a manner that is reflective of the demands placed on the system by various groups of customers. As shown in Figure 1.1, the basic elements of cost-of-service rate-making include the following:

- Determine revenue requirement,
- Categorize customers into different customer classes,
- Functionalization of costs,
- Allocate utility costs to cost-causative components,

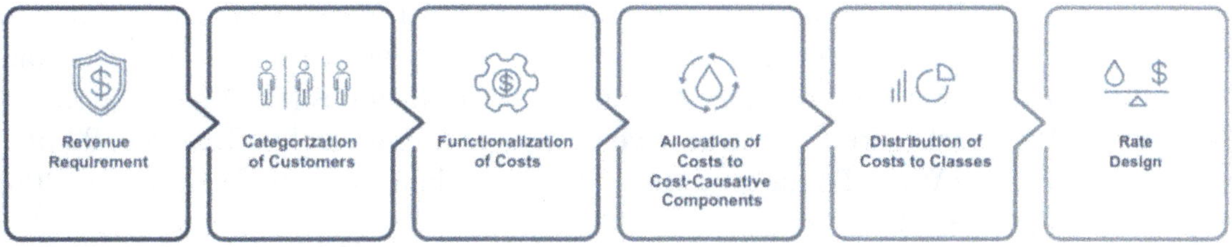

FIGURE 1.1 Cost-of-Service Rate-Making Process

- Distribute costs to customer classes based on estimated flows and strength constituents, and
- Design a structure of rates and charges that will recover allocated costs from each customer class.

2.0 HISTORICAL OVERVIEW

Efforts have been made to formulate and apply rational and consistent principles for allocating utility costs to develop a basis for fair wastewater service rates. Such principles were clearly stated for the first time in 1951 by a committee representing many diverse disciplines, including engineering, law, accounting, and finance (Ohio State Law Journal, 1951). The basic costing criterion advocated by that committee was that

> The needed total annual revenue requirements of . . . sewage works shall be contributed by users and nonusers (or by users and properties) for *whose use, need and benefit the facilities of the works are provided, approximately in proportion to the cost of providing the use and the benefits of the works* [emphasis added].

The committee's principle of fair sharing may, in retrospect, seem obvious; however, it was a genuine departure from common practice at the time. Even in 1973, when the principle was reaffirmed by a similar committee, more than half of the wastewater systems in the United States obtained their entire revenue requirements from property taxes.

3.0 WATER POLLUTION CONTROL ACT, AS AMENDED

Wastewater utility rate setting is unique, partly because, for many utilities, a substantial portion of their rate structure must comply with federal regulations. As part of the United States' historic 1972 Federal Water Pollution

Control Act amendments, commonly referred to as the Clean Water Act, U.S. Congress funded a construction grants program to help utilities construct the necessary water resource recovery facilities (WRRFs) to meet the Act's goals. However, to receive these grant funds, Public Law No. 92-500 (Federal Water Pollution Control Act Amendments of 1972), in Section 204, (b) (1), required that

> Notwithstanding any other provision of this title, the Administrator shall not approve any grant for any treatment works . . . unless he shall first have determined that the applicant (A) has adopted or will adopt a system of charges to assure that each recipient of waste treatment services within the applicant's jurisdiction . . . will pay its proportionate share of the costs of operation and maintenance (including replacement) of any waste treatment services provided by the applicant; (B) has made provision for the payment to such applicant by the industrial users of the treatment works, of that portion of the cost of construction of such treatment works . . . which is allocable to the treatment of such industrial wastes . . .; and (C) has legal, institutional, managerial and financial capability to insure adequate construction, operation and maintenance of treatment works throughout the applicant's jurisdiction . . .

This was not the first time that a federal directive addressed the issue of rate-setting (Pub. L. No. 84-660 [amendments to the Federal Water Pollution Control Act, 1956] also did, but to a much lesser extent); however, it was unique in this case because of the size of the construction grants program (which meant that most utilities would be affected by this provision) and the extent to which this provision, known as the user charge provision of the Act, affected rate setting. The scope of this section was considered necessary by legislators because few utilities recovered costs through fair and equitable rates and, instead, recovered their costs through the jurisdiction's ad valorem tax system. These provisions were considered essential to preserving the investment being made with federal funds provided under the Act.

Over time, there were changes and clarifications to this section of Pub. L. No. 92-500, as follows:

- In the 1977 amendments, a limited number of utilities were allowed to continue using ad valorem taxes to pay for wastewater treatment.
- In guidance later provided by the U.S. Environmental Protection Agency (U.S. EPA), the meaning of the term *including replacement* was clarified by renaming it *equipment replacement*.

- Condition B, the industrial cost recovery provision, was repealed in 1981.
- New amendments, such as the industrial pretreatment regulations, produced new charges.

A careful reading of Section 204 defines some of the requirements of a user charge system as follows:

- The utility must know its operations, maintenance, and equipment replacement costs. (Although Section 204 did not address capital costs, a sound utility rate structure should recover capital costs from its users.)
- The rate methodology must be proportionate. This means that unit rates should be the same regardless of the volume used (known by some as "uniform" rates). (However, the lack of specificity of types of costs and levels of service can potentially lead to nonproportionate rates. For example, because Section 204 did not address capital costs, it is possible for a nonproportionate rate structure to meet this requirement.)
- Proportionality requires that the rate-setting entity must have a basis for charging. ("A system of charges . . . may be based on something other than metering the sewage or water supply flow of residential recipients of waste treatment services, including ad valorem taxes. If the system of charges is based on something other than metering, the Administrator shall require [a] the applicant to establish a system by which the necessary funds will be available for the proper operation and maintenance of the treatment works; and [b] the applicant to establish a procedure under which the residential user will be notified as to that portion of his total payment which will be allocated to the costs of the waste treatment services." [Federal Water Pollution Control Act of 1956]). This can be challenging because not all water passing through a water meter is returned to the WRRF.
- Proportionality also requires that users with higher-than-normal pollutant strength discharges pay more for this strength handling.

4.0 MEASUREMENT OF BENEFIT AND USE

Wastewater service is a critical service, or function, that contributes to and sustains a safe, clean environment. However, customers' wastewater can vary greatly in pollutant strength and quantity. Some wastewater may be of nearly tap water quality, while other discharge sources, such as septic tank sludge, may contain high concentrations of organic matter. Typically, several significant types of pollutants are considered in determining wastewater strength and associated treatment processes: one is for biodegradable organic matter, and the other is for suspended solids that can be settled from the water. For some wastewater systems, other specific contaminant removal may also be important. For example, using synthetic oils from industry can greatly increase the chemical oxygen demand levels and the treatment requirements for that effluent. Likewise, excessive nutrient discharges can have detrimental effects on receiving bodies and must be remediated.

Wastewater flow measurement for small customer accounts is impractical. Thus, water meters are generally relied on to estimate contributed flows to the wastewater system. Often, large volumes of water are used for lawn irrigation, car washing, and in processes involving evaporation and other consumptive uses (cooling water operations). Typically, such water uses are not returned to the wastewater system. Accordingly, a fair rate should recognize each system's water and wastewater volume differential.

Another complication is that the total flow reaching many collection systems and WRRFs includes a substantial amount of groundwater infiltration to pipes and manholes and inflow of extraneous surface runoff. Most inflow and infiltration are difficult to prevent. Also, many older combined sewer systems collect and convey stormwater inflows from street gutters, roof drains, residential basement sump pumps, and manholes. Nonetheless, this additional flow still must be transported, pumped, and processed with sanitary sewer discharges.

Peak flow variability is a complex factor for flow handling and treatment requirements. Wastewater flows are extremely variable between peak and average flows, partly because of infiltration and inflow, as discussed above. A wastewater utility can sometimes experience a peak flow many times greater than its average, which creates several problems for the wastewater collection and treatment process. Management of peak flow represents a significant cost for utilities and must be incorporated to the rate structure.

Given these complexities, establishing cost responsibility (and ultimately cost-of-service rates) must rely on practical approaches to cost allocation, allowing for recognition of individual cost-causative components. Costs must be allocated among average and peak flow handling, pollutant removal (including biochemical oxygen demand, suspended solids, and nutrients), and customer billing and administration so that costs can be allocated to customer classes in a manner that is proportionate to each class's use of the system. The principle is more straightforward for residential, commercial, and industrial discharges. Assignment of costs related to the conveyance and treatment of stormwater inflows or groundwater infiltration and the recognition of benefits for future availability of wastewater service (unused or excess capacity and service availability) provide greater challenges in the rate-making process.

The cost-allocation techniques presented in this MOP generally apply to most wastewater utilities. Basic guidance is provided without regard to whether a supplier is under the jurisdiction of any federal, state, or local agency. The provisions presented are also generally considered consistent with interpretations of the law for receiving grants through U.S. EPA.

5.0 MANUAL OF PRACTICE CHAPTER CONTENTS

This MOP is structured to provide the reader with an overview of financing mechanisms for wastewater systems and a framework for developing wastewater rates and charges.

CHAPTER 2. The spectrum of available institutional structures used to deliver wastewater services, ranging from multiparty agreements to various forms of districts and authorities, to public–private partnerships, are outlined. In addition, legal issues affecting the development of wastewater user charges, including federal user charge regulations and state and local legal considerations, are discussed.

CHAPTER 3. Financial management and accounting practices necessary for the successful operation of a wastewater utility, including financial planning, budgeting, cost accounting and reporting, and establishment of financial reserves, are discussed.

CHAPTER 4. The special issues involved in financing capital improvements are discussed, including the capital planning process, internal and external sources of funds, investor-owned utility capital financing, and the development of capital financing policies.

CHAPTER 5. The fundamental steps involved in the development of cost-of-service-based rates begin with the determination of revenue requirements or the overall amount of revenues that must be generated from the utility's rates and charges. This chapter, along with an accompanying Appendix A, provides discussions and numerical examples of the determination of revenue requirements using the "cash-basis" and "utility-basis" approaches.

CHAPTER 6. The two accepted methodologies for allocating annual revenue requirements to cost components are presented. Design-basis and functional cost-allocation methodologies are described, their relative advantages and limitations are discussed, and numerical examples are provided.

CHAPTER 7. The definition of customer classes, the process of allocating cost components to service characteristics (including flows, strength loadings, and customer metering and billing), and the distribution of costs to defined customer classes are discussed, and numerical examples are provided. In addition, a facility-balance analysis that provides an accounting of system flow and strength contributions is outlined.

CHAPTER 8. The "art" of rate design is discussed, including a review of rate design policy objectives, a review of alternative rate forms and structural components, and the introduction of a rate evaluation matrix. Approaches for determining billing units are discussed. Also, calculations of rates and associated customer bill effects under systemwide and cost-of-service-based uniform rates by customer class are presented.

CHAPTER 9. Wet weather financing and cost recovery have become a significant part of wastewater finance and pricing. Whether collection systems are separate or combined, stormwater inflow and groundwater infiltration contribute heavily to system costs. This chapter identifies methods for identifying, allocating, and recovering these costs from customers.

CHAPTER 10. System development charges (or impact fees) represent an important mechanism by which utilities may recover costs associated with ensuring capacity availability to serve new users to the utility system. This chapter reviews the legal considerations involved in imposing system development charges, alternative methods of calculating such fees, and associated program evaluation and monitoring requirements.

CHAPTER 11. Given that any modification to a utility's system of rates and charges will significantly affect a broad spectrum of stakeholder groups, careful planning for implementing such changes is critical. This chapter outlines the fundamentals of effective implementation planning and provides guidance on how such planning may be tailored to the needs of individual communities.

CHAPTER 12. This chapter discusses the affordability of prospective utility rates. Affordability standards and thresholds promulgated by U.S. EPA are discussed in addition to opportunities to enhance customer affordability via alternative rate designs or programs.

CHAPTER 13. Developing rates for reclaimed water presents several unique issues and challenges in that the water that is of lower quality but provides nutrient value, it is a means of wastewater effluent disposal, and it may mitigate system peaking factors. These issues are outlined in addition to the process for cost allocation, rate design, and determination of system development charges for reclaimed water.

CHAPTER 14. As the costs of wastewater management rise, utilities experience increasing pressure to effectively fund capital improvements, which has led the industry to identify new, internal financing alternatives. When used in conjunction with conventional financing, these alternatives result in more efficient paths to achieving the utility's capital objectives.

APPENDIX A. Appendix A presents the "utility-basis" approach to determining revenue requirements and allocating costs by cost-causative components to customer classes using the design-basis cost-allocation methodology.

6.0 UNITS OF MEASUREMENT

Typically, wastewater rates are reported in cost per unit of volume. In the United States, utilities are seemingly evenly split between expressing volumetric rates in terms of cubic feet or gallons (1 Ccf = 2.83 m^3). This MOP notes rates as a cost per hundred cubic feet ($/Ccf). Because there are 748 gal in 100 cu ft, $/Ccf may be converted to $/1000 gal (kgal) by dividing by 0.748. For example, a rate of $3.00/Ccf equals $4.01/kgal ($3.00/Ccf divided by 0.748 kgal per Ccf = $4.01/kgal).

7.0 REFERENCES

Economic Development Research Group for American Society of Civil Engineers. (2016). *Failure to act: Closing the infrastructure gap for America's future*. American Society of Civil Engineers.

Federal Water Pollution Control Act Amendments of 1972, 33 U.S.C. §1251 et seq. (1972).

Federal Water Pollution Control Act of 1956, Pub. L. No. 84-660.

Ohio State Law Journal. (1951). *Joint committee report, 12,* 151.

2

Institutional Framework for Wastewater Services

1.0 INTRODUCTION

Every user charge system is influenced by a variety of fundamental institutional issues. Some of these issues are directly relevant to the calculation of user charges; however, others, while less directly connected, still have important implications. This chapter introduces some of the most important institutional factors that managers and rate practitioners need to recognize when developing sound and implementable user charge systems. For purposes of this chapter, institutional issues include (1) external organizational structure, including the legal environment and forms of governance; (2) internal organizational considerations; (3) other institutional agreements and interactions with a bearing on utility operations, management, and cost recovery; and (4) governmental regulations.

Governance generally refers to the structure by which the utility is legally organized. Specifically, it is the legal form by which the utility performs its daily responsibilities and relates to its stakeholders (customers, regulatory agencies, developers, other governments, lending institutions, suppliers, advisors, etc.). Institutional forms that will be discussed in this chapter include the following:

- Single community,
- Independent authority,
- Special purpose district,
- Multipurpose district,
- Joint ownership,
- Regional agencies, and
- Private ownership.

The categories are complex and interrelated, and many variations on the aforementioned structures are in operation around the country, including the presence or absence of enterprise funds within these institutional forms. An example of additional complexity is that privately owned and operated utilities, while facing some different institutional considerations than those facing public agencies, may operate in patterns similar to other institutional environments listed above.

Internal organizational considerations include utility-specific factors related to daily operations that affect the establishment, review, and implementation of user charges. These organization factors include the following:

- System operations,
- Staffing and data management, and
- Utility systems.

Other institutional considerations also deserve to be mentioned because agreements and other external interactions may affect user charge development. These include the following:

- Lender requirements, including bond covenants;
- Multiparty agreements;

- Contracts for service outside of a local jurisdiction;
- Special service contracts and user agreements; and
- Private ownership and public–private partnerships.

Governmental regulations include federal, state, and local legal requirements with which utilities must comply. These requirements would include the following:

- Federal user charge regulations as promulgated by the U.S. Environmental Protection Agency (U.S. EPA), especially those related to the development of wastewater rates and charges;
- State requirements developed by state environmental agencies and lending institutions (e.g., state revolving funds [SRFs] or other state loan programs), and public service or public utility commissions for private utilities;
- Local and regional laws and ordinances that define how wastewater costs must be recovered from all customers or specific classes of customers (governmental customers, outside city customers, wholesale customers, etc.); and
- Other specific regulations or laws with which a utility must comply.

2.0 INSTITUTIONAL ORGANIZATION

An overall discussion of the financial aspects of wastewater service would be incomplete without a brief consideration of the broad institutional alternatives for the administration of this service. In most cases in the United States, the responsibility for wastewater and stormwater treatment and conveyance rests with bodies of elected or appointed officials, ranging in scope from a semiofficial subdivision development board to the multiple-function metropolitan authority. Some of these bodies exist solely for the purpose of administering pollution control and wet weather management facilities, while many are local governments providing multiple types of infrastructure and services. Some of the more common forms of organizations are single community, independent authority, special purpose district, multipurpose district, joint ownership, regional agencies, and private ownership. Even though managers, leaders, and rate analysts within existing organizations may find it challenging to change their organizational structure, a brief review of different institutional arrangements can offer valuable perspectives. The comments in the following subsections do not aim to provide a definitive guide to organizational forms but rather to highlight potentially relevant and interesting aspects for the reader.

2.1 Single Community

The most common method of providing service is single community ownership, operation, and administration of the water resource recovery facilities (WRRFs) where the conveyance, treatment, and disposal of clean effluent and biosolids or residuals is considered a municipal service to the community's residents.

The governing bodies of many communities exercise the administration of wastewater systems directly, through the municipal public works department, a separate wastewater or utility department, or sometimes through an engineering department. In some communities, administrative responsibilities rest with an appointed board, commission, or similar body whose functions also may include the provision of other municipal drainage facilities. Generally, in the single community institutional form, the elected municipal officials retain the ultimate responsibility for policy decisions and utility performance even though the appointed board exercises much of the decision-making.

In some instances, particularly in large communities or certain states, a separate body is elected or appointed to exercise control over wastewater treatment, either alone or with other functions, even if a service area is solely within the limits of one municipality.

2.2 Independent Authority

Independent wastewater authorities are often organized to create a single-purpose entity to construct and operate public WRRFs, sometimes combined with water and stormwater services. An authority may have geographical limitations, overlap a number of local jurisdictions, and even cross state lines.

Most authorities do not have taxing power, so they tend to be largely reliant on revenues from rates and charges. The smaller and less diverse revenue stream of an authority might result in slightly less advantageous financing terms relative to general governments with taxing power, but municipal finance markets recognize the stability and strength of well-managed authorities, so financing issues need not arise. Authorities are generally financially independent from any other entity, which can be beneficial in avoiding the burdens of nonutility financial management issues in a service area.

2.3 Special Purpose District

Pragmatic and cooperative solutions to wastewater management problems sometimes lead to the creation of a special purpose district. Special purpose districts are similar to other nonmunicipal forms of organization but can sometimes be characterized as creating more power for the entity to combine resources and providing planning and management services for the communities involved. In some states, such communities must be in the same county; in other states, a special legislative action must incorporate the district as a separate governmental unit. Enabling legislation regarding special purpose districts varies widely; some legislation allows the creation of districts with strong powers, while other legislation significantly limits district authority. Districts can be focused on retail service, without regard to other local governments in the area, or they can be focused on the cities, counties, or other entities involved.

The special purpose district can be particularly relevant to situations in which large capital expenditures are required for joint projects. For "entity focused" districts, an elected or appointed board representative of the participating communities is typically established that has the authority to levy or collect charges as required to finance the district's needs. Debt undertaken to finance capital improvements can be a full-faith and credit obligation of all the communities involved, even though the bonds are issued in the name of the district and are serviced through district revenues. Some districts do not issue bonds directly; instead, the member communities finance their district obligations directly, as in the joint ownership arrangement.

A district serving multiple participating communities might be focused just on providing joint conveyance and WRRFs used by multiple communities, while member communities continue to construct and operate their own independent facilities. Special purpose districts may be particularly valuable in facilitating cooperative efforts toward cost-effective wastewater management where multiple and varied jurisdictions are involved.

2.4 Multipurpose District

The organization and functioning of a multipurpose district are similar to that of a special purpose district. A multipurpose district is typically formed to approach multiple intergovernmental problems such as water pollution control, water supply, irrigation, solid waste disposal, flood control, fire protection, and navigation. Typically, areas affected by water pollution control problems are in need of community cooperation in other matters as well; this is particularly true in metropolitan areas.

One of the challenges of a multipurpose district is the delineation of district boundaries to include those communities having common problems without excluding communities in need of a cooperative approach on a limited number of problems. If, for example, the district was to function in the areas of water pollution control and flood control, the communities situated on higher ground would need a WRRF located in a community closer to a significant receiving stream located on lower ground but would not necessarily need flood control.

One advantage of a multipurpose district compared to a special purpose arrangement is the potential economic advantage of using facilities, professional staff, and other resources for a number of similar problems over a greater area. Furthermore, multipurpose districts often meet their revenue requirements through a combination of assessments made on the district as a whole and through the application of service charges for the various services provided by the district.

2.5 Joint Ownership

Two or more communities or a community and another entity (such as a governmental agency or large industrial firm) may construct and jointly operate WRRFs. In organizing joint wastewater activities, the individual local governing bodies define the procedures for administration and financing and typically create a joint body composed of elected or appointed officials from the cooperating governmental or other units.

Joint ownership of facilities is typically considered to differ from the district approach because the vested title remains with the individual units in proportion to their expenditures of capital or to the cooperating communities as joint tenants. Likewise, it would be typical for indebtedness to be incurred by individual communities rather than by the joint body created.

Methods of cost distribution among the participants are typically defined in a joint agreement and, while these methods can vary significantly, proportionality to capacity demands or usage patterns provides a common thread. Joint bodies rarely assess and collect the charges from retail customers directly but rather apportion the costs to the cooperating participants who, in turn, recover these costs from their retail customers. The methods of assessment and collection used by the parties need not be the same.

Joint ownership is particularly advantageous in the case of a small number of communities located adjacent to or close to one another to take advantage of the lower cost of construction and operation of a joint WRRF or a jointly used conveyance system. If a relatively large number of communities are involved, questions of central administrative authority, adding service area, and transfer of capacity rights can be challenging.

2.6 Regional Agencies

Regional governmental agencies may be created to handle the administration and operation of wastewater conveyance and treatment. A number of these agencies are created to provide comprehensive regional water, wastewater, and stormwater planning, while others have been formed exclusively for water pollution control. The regional agency can take on the characteristics of an independent authority, special purpose district, or multipurpose district. Economies of scale may also provide an impetus for the creation of such multipurpose agencies. These agencies typically administer the program for a metropolitan area, river basin, or other geographically delineated area such as a regional watershed. Legislative action may be required to initiate this type of arrangement.

The most obvious advantages are the centralized pollution control ability of such an arrangement, standardization of equipment and procedures, the maintenance of highly trained personnel to provide expertise on operating problems and procedures, centralized laboratory facilities, and the coordination of water quality monitoring with facility performance.

2.7 Private Ownership

Privately owned companies may purchase previously publicly owned utilities or establish a new utility to provide wastewater services and can serve within a single jurisdiction or across jurisdictional boundaries. Private utilities are typically subject to rate regulation by a state public utilities commission or public service commission and are subject to environmental regulations affecting public agencies.

One reason that private ownership is sometimes instituted is that owners are typically given authority to establish user charges sufficient to meet the financial needs of the utility in a way that is somewhat insulated from local political or external pressures. For private utilities serving multiple service areas, an additional advantage may arise from the application of specialized personnel or other resources over a larger customer base.

One challenge for private utilities can be the perceived lack of public involvement or influence on the development of user charges that are approved at either a regional, state, or provincial level and whose ownership may not be local to the system's users.

3.0 SELECTING THE APPROPRIATE INSTITUTIONAL ORGANIZATION

The institutional arrangements outlined above are not the only means of providing for wastewater service, but they are applicable in most cases. Other arrangements, as dictated or modified by local laws and unique situations, can provide methods for such administration. When determining which type of arrangement is most beneficial for a particular situation, several factors must be considered. Some of these are addressed under the descriptions of each institutional arrangement, but areas deserving of particular focus include the following:

- Technical and financial capabilities of various participants,
- Historical and current relationships among participants,
- Risk tolerance of participating entities,
- Achievement of the benefits of economies of scale,
- Effect on costs and user charges,
- Drainage basin and geographic considerations,
- Wastewater influent characteristics and statutory wastewater effluent permit criteria,
- Location of important resources,
- Acceptance by customers being served,
- The perspective of financial markets and bond-rating agencies, and
- Acceptance by state environmental agencies.

Often, these factors conflict and a careful evaluation of various tradeoffs must be performed before selecting an institutional framework.

4.0 INTERNAL ORGANIZATIONAL CONSIDERATIONS

A utility's internal organization is relevant to the effective development of user charges. The legal structure creating a utility and the external institutional structure within which a utility exists were addressed previously. This section mentions several internal organizational considerations that deserve attention. Although external considerations may be preexisting and represent unchangeable or at least inflexible conditions for a manager, internal considerations may provide greater latitude within which managers can beneficially influence the conditions relevant to establishing or refining a user charge system.

4.1 System Operations

Utility systems are heavily affected by specific external factors related to geography and system configuration, customer base, and applicable effluent discharge limitations. Different types of treatment processes and differing levels of complexity in collection systems have implications for the development of rates and charges. A modestly sized wastewater utility with a relatively homogeneous customer base is not likely to need to consider the breadth of issues arising from more complex treatment issues or elaborate rates and surcharges for highly variable or particularly challenging high-strength wastes. Similarly, utilities producing and selling reclaimed water may encounter significantly increased complexity in pricing and management because of inherently dual-purpose aspects of reuse, overarching policy drivers, and as-yet immature markets for reuse water in many cases.

4.2 Staffing and Data Availability

Establishing, reviewing, and/or implementing a user charge system requires a significant amount of data in addition to the investment of staff resources at multiple levels. As such, determining the responsibilities of staff to perform the analytics, management, and implementation of user charges should be a primary consideration for management. Objectives should be outlined, required and available data identified, and an implementation strategy established as part of any review or analysis of rates.

4.3 Utility Billing Systems

Modern utilities increasingly rely on complex technology, computer operating systems, and customer-facing applications to provide wastewater services to customers. Implementation of user charges must be made in conjunction with the available systems that the utility uses to provide, measure, and bill for such services. In addition, customer transparency and dramatically increased interaction is increasingly being requested and expected. The challenge of these new expectations provides an opportunity for management as it relates to communicating user charge study processes and outcomes. An additional challenge may be faced by single-purpose wastewater utilities that are potentially dependent on billing systems and data under the management of others.

5.0 OTHER INSTITUTIONAL CONSIDERATIONS

In addition to institutional organizational forms and internal organizational structures, contracts and financial obligations play an important role in providing wastewater service. Public–private partnerships have also become more widespread as alternatives to traditional wastewater models. These concepts are introduced below.

5.1 Lender Requirements

Utility systems are capital intensive and, often, are financed through a variety of debt instruments. When acquiring a debt instrument, utilities typically enter into legally binding covenants with the bond or debt issuer. Clauses within these agreements often dictate operational requirements, accounting methods, system connections, levels or user charges, collection processes, user charge service provisions, and other factors.

Specific to user charges, these documents often include statements that the debt issuer may not render free services and may not reduce user charges unless adequate revenues are projected. In addition, revenues recovered from user charges must meet stated required levels to meet debt service requirements.

Lender agreements are important for both the debt holder and the debt issuer and management must be aware of the individual provisions of associated requirements.

5.2 Multiparty Agreements

Several of the institutional alternatives discussed in this chapter require multiparty agreements. The function of a multiparty agreement is to establish mutually dependent promises of service obligations and responsibilities, both financial and performance related, that serve to govern the relationship of the parties. The agreement should address both current and future conditions. From an implementation perspective, the executed multiparty agreement serves as a notice to the financial community that the agency has a revenue source to satisfy its debt service obligation and that the parties are required to make the payments necessary to ensure that the agency meets its financial obligations.

The terms of the agreement are governed partly by regulatory requirements and partly by practical considerations. Issues and concerns can be divided into three broad categories: construction of significant projects, user charges, and miscellaneous.

The most important aspect of project construction is a pledge to complete the necessary planning and construction because a party is entrusting its wastewater service responsibilities to another party. In return, the other party must pledge to deliver its wastewater to the treating agency and to pay for appropriate costs.

The multiparty agreement must set forth the responsibility and method of calculating wastewater user charges. Because the treating agency will typically need revenue at project startup and a user may have a need for service by a particular date, it is critical that the agreement contain a schedule to establish the date when payments, associated revenue requirements, and services are to begin. This need for revenue and service should supersede any individual party's desire to delay payment or service obligations. The agreement should also require the continuation of payments and service during negotiations or litigation among the parties so that the agency can continue to function financially and the user can obtain wastewater service.

Miscellaneous issues include (1) parallel adoption of rules and regulations; (2) degree of pretreatment required, including categorical and local-limit requirements; (3) inflow and infiltration control; (4) duration of agreement and effects of breach of agreement; (5) covenant against noncompeting facilities; (6) cost-sharing requirements for ongoing operation and maintenance of wastewater collection and treatment infrastructure and for addressing future capital needs; (7) procedures for resolution of disputes; and (8) allocation of available capacity.

5.3 Contract for Service Outside of Local Jurisdiction

In cases where, for example, urban facilities are asked to serve suburban development, a contract can provide the most acceptable approach. The contract approach allows for intercommunity cooperation on either a permanent or a temporary basis without the necessity of reorganizing the original service entity. The key is mutual agreement on the terms of a contract that clearly define the quantity and standards of service; magnitude of charges, provisions for future capital, if needed; responsibilities and liabilities of the contracting parties; duration of the agreement; and provisions for renegotiation.

Such contracts can be particularly useful for the temporary provision of service to a growing community until development requires the provision of local WRRFs. In addition, they can provide sustainable and effective long-term arrangements between communities and sanitary or sanitation districts. Contracts can also be advantageous in a municipal–industrial cooperation for the construction of WRRFs; however, such contracts may need to stipulate guaranteed levels of minimum revenue from industrial partners to justify utility investment or use of existing capacity.

Contracts for service can be seen as having disadvantages regarding representation. Typically, the secondary contracting party is not involved in the policy decisions of the entity providing service, even though such decisions may affect the secondary party as well. A possible solution is to define certain rights, responsibilities, and requirements for input from the secondary party in the agreement to the operator so that such policy decisions reflect consideration of the secondary party's interests and needs. Despite these shortcomings, the contract for service has become a common and useful means for providing interentity cooperation.

5.4 Special Service Contracts and User Agreements

Even in cases where neither multiple utilities nor jurisdictional boundaries are involved, utilities can sometimes use special service contracts or user agreements. Such agreements can address wastewater treatment requirements for unique or unusual waste streams (such as landfill leachate or hydrofracturing wastewater), location or other service characteristics significantly at variance with typical customer requirements, or special considerations associated with unusually large customer demands.

5.5 Privately Owned Utilities and Public–Private Partnerships

As indicated in Section 2.7, private ownership of utilities is an institutional alternative for providing wastewater services. Similarly, public–private partnerships may also provide for private-sector ownership or operations of all or selected wastewater services; the variety of arrangements is substantial (for example, design–build–operate, design–build–operate–maintain, design–build–own–operate–maintain), but the concept of contracting for services is generally applicable. Privately owned utilities for full wastewater operations are less common than for water operations, reflecting complex wastewater regulations and pre-existing institutional arrangements. On the other hand, it is common to see private contractors involved in a plethora of wastewater activities ranging from contract operations to design–build–operate facilities.

6.0 SELECT GOVERNMENTAL REGULATIONS

6.1 Clean Water Act User Charge Regulations

The original Federal Water Pollution Control Act Amendments of 1972 (commonly known as the Clean Water Act [CWA]) and its subsequent amendments provide for the regulation of certain financial aspects of wastewater systems, including the development of user charges. (The term *user charge* has historically referred to the total charges to a user whether for debt service, operation, cash-financed improvements, or maintenance. U.S. EPA used the term in this context to mean the operation, maintenance, and equipment replacement costs of a wastewater system.) A majority of publicly owned wastewater utilities in the United States availed themselves of federal grants in the years immediately following the CWA, thereby becoming subject to user charge regulations. While these user charge requirements originally were relevant only for systems that received such grants, to a large extent they have become a part of industry-accepted guidelines for user charge ratemaking practices. Additionally, after the expiration of these grant programs, these regulations, or at least the direction established by these regulations, have been applied to the administration of SRF loans where federal funds were involved, among other circumstances, including being reflected in user charge regulations beyond the United States. Thus, CWA user charge guidelines continue to be relevant more than four decades later.

The CWA regulations include project-specific requirements, management issues, and rate ordinance issues. Project-specific requirements relate primarily to treatment technologies and do not directly affect the structure of a user charge system. Management-related sections of CWA require that grant and loan recipients demonstrate an appropriate level of managerial competence. This demonstration includes, but is not limited to, thorough explanations of (1) the responsibilities of all involved government entities, (2) the methods used to account for and finance all future costs related to the grant-financed facilities, and (3) a calculation of the annual cost of wastewater management per household. Furthermore, utilities are required to ensure an adequate level of public participation through public notices and hearings. Although not directly relevant to the calculation of rates and charges, these provisions do reveal an intent to support the creation and ongoing operation of well-managed utilities.

Rate ordinance issues are the key focus of this manual of practice (MOP), and the key focus of CWA rate requirements is on an equitable allocation of wastewater operations, maintenance, and replacement costs to a series of appropriately defined user classes.

U.S. EPA-recommended methodology for constructing an equitable user charge ordinance is provided in their publication, *User Charge Guidance Manual for Publicly-Owned Treatment Works* (U.S. EPA, 1984). This methodology consists of identifying and functionalizing system costs, identifying and analyzing customer classes, and allocating costs to the appropriate customer class in proportion to the class's use of the system. The majority of this MOP is devoted to an exposition of these cost-of-service steps. U.S. EPA's user charge manual also requires that utilities be able to demonstrate the ordinance's ability to generate sufficient revenue to "pay the total operation and maintenance costs necessary to the proper operation and maintenance (including replacement) of the treatment works" (Federal Water Pollution Control Act Amendments of 1972). These two fundamental requirements, equity in cost allocation and revenue sufficiency, provide a firm foundation for the development of user charges today.

Programs to address low-income/affordability concerns were recognized through an amendment to the CWA (Federal Water Pollution Control Act Amendments of 1972), which introduces the concept of low-income subsidization or discounted rates to eligible residential customers. Although the subsidization of low-income users would violate fundamental proportionality requirements, the amendment provides that "A system of user charges which imposes a lower charge for low-income residential users shall be deemed to be a user charge system meeting the requirements . . . if the Administrator determines that such system was adopted after public notice and hearing" (Federal Water Pollution Control Act Amendments of 1972). This provision should not be considered a blanket authorization of any potential or specific program but, subject to other applicable rate-setting objectives and guidance, affordability considerations may be found to be consistent with CWA requirements. U.S. EPA's guidance continues to evolve as affordability issues are being addressed in different ways in different communities. The most current guidance as of this writing is provided in *Clean Water Act Financial Capability Assessment Guidance* (U.S. EPA, 2024).

The CWA also requires grant recipients subject to CWA regulations to require wholesale customers with their own retail customers to establish retail user charge systems consistent with CWA user charge requirements.

Financial reporting requirements are also applicable to utilities receiving federal or state grants and loans (or other debt proceeds), and these reporting requirements often require that financial records be submitted to document the financial health of the utility. Specific requirements vary, but the general requirement to document the financial health of the utility is consistent with the fundamental tenets for development of user charges and operation of a financially sound utility.

6.2 Other Governmental Regulations and Issues

Wastewater utilities are also subject to a variety of state and provincial regulations and, possibly, even local regulations potentially applicable to the development of user charge systems. Consistent with laws applicable to local governments generally, utilities are often granted pricing and rate-setting authority by specific legislation or are said to have such authority by virtue of their obligation to meet the public health and environmental needs of their communities. It is important for any utility manager or rate practitioner to understand the specific state/provincial and local regulatory environment applicable in their situation.

One increasingly important issue in recent years is the question of addressing the ability of lower income customers to afford their water and sewer bills. Federal and state regulations such as CWA or California's Proposition 218 (known as the Right to Vote on Taxes Act of 1996) that stipulate, limit, or prohibit programs that provide intercustomer subsidies or discounts to certain customers must be considered. Concerns over a range of social topics are becoming a part of rate conversations in some communities, necessitating additional analyses of distributive or equity considerations to address environmental justice and related topics.

The question of how to pay for necessary assets and services in light of customer assistance programs without violating restrictions on rate setting and the appropriate use of rate revenues is receiving increasing attention, including in *Drinking Water and Wastewater Utility Customer Assistance Programs* (U.S. EPA, 2016), which was created by U.S. EPA in cooperation with the Water Environment Federation and other significant water industry associations. This compendium was followed by *Navigating Legal Pathways to Rate-Funded Customer Assistance Programs: A Guide for Water and Wastewater Utilities* (University of North Carolina Environmental Finance Center, 2017), which documents legal constraints to the development and funding of customer assistance programs facing utilities in all 50 U.S. states. State- and locality-specific conditions dominate, with U.S. or industry-wide guidance having yet to emerge. Additional programs providing technical assistance and access to funding continue to be developed, including those addressed in U.S. EPA's *Closing America's Wastewater Access Gap* initiative (https://www.epa.gov/water-infrastructure/closing-americas-wastewater-access-gap).

7.0 REFERENCES

Federal Water Pollution Control Act Amendments of 1972. 33 U.S.C. §1251 et seq. (1972).

U.S. Environmental Protection Agency. (1984). *User charge guidance manual for publicly-owned treatment works* (EPA 430/9-84-006). Office of Water Program Operations.

U.S. Environmental Protection Agency. (2016). *Drinking water and wastewater utility customer assistance programs.*

U.S. Environmental Protection Agency. (2024, October 1). Clean Water Act financial capability assessment guidance. https://www.epa.gov/waterfinancecenter/clean-water-act-financial-capability-assessment-guidance

University of North Carolina Environmental Finance Center. (2017). *Navigating legal pathways to rate-funded customer assistance programs: A guide for water and wastewater utilities.*

3

Financial Management and Accounting for Wastewater Systems

1.0 INTRODUCTION

The concept of sustainability extends beyond engineering to the financial policies, management, and accounting systems that support the activities and operations discussed later in this manual. Additionally, increasing public scrutiny and calls for transparency mean that utilities must now prove they are good stewards of ratepayers' money. This chapter focuses on policy development, financial planning, budgeting, cost accounting and financial reporting, and the need for and use of reserves to manage financial and operational risks effectively. Chapter 3 provides an overview of these topics in broad terms; where appropriate, the chapter provides references for additional information on each subcategory. The information in this chapter aids in developing financial information and reporting systems that can provide data necessary to establish financially sustainable practices and cost-based rates and charges.

2.0 POLICY DEVELOPMENT AND ADOPTION

The first step in establishing a clear and supportable financial-management program for all levels of an organization involves policy development and adoption. Utility executive management must establish comprehensive guidelines that align with the needs of various organizational levels and clearly define the utility's financial and accounting operations. These policies should support measurable outcomes aligned with the utility's strategic goals. To ensure that the utility manages ratepayer funds responsibly, effectively, and efficiently, utility management should provide high-level information demonstrating goal achievement and sufficient controls over revenues and expenditures. It is important for management to have access to year-to-date expenditure levels so that they can determine where the utility and its departments stand financially with respect to managing its allocations for specific programs and activities. Finally, those individuals that have direct responsibility for purchasing, contract administration, service contracting, and maintenance need to know that they will have the resources and financial means to support day-to-day activities. All levels of an organization should understand and operate using the same consistent and coordinated policies and procedures.

The laws and regulations that underlie the governance of the utility and the basic operating philosophies related to utility services may affect the definition of the types of policies and processes requiring definition. These might include the following:

- Operating, resiliency, and capital reserve levels;
- Financial targets, such as debt-service coverage, debt-to-equity ratios, liquidity, and capital re-investment;
- Policies regarding the administration of risk and insurance for the utility;
- Identification of the means and methods of accounting for revenue and expenses; and
- Transparency regarding the utility's financial operations.

From a rate-setting perspective, utilities may also consider developing additional financial policies addressing concerns such as the following:

- Maintaining/achieving revenue stability;
- Prioritizing conservation efforts, such as development of reclaimed water as an alternative water source and revenue stream; and
- Assessing the need for affordability programs.

Some of the aforementioned policies may present conflicting objectives and, therefore, clear policy guidance is important for all professionals within a utility.

These polices should clearly present the goals of the utility's decision-making body when it comes to financial and fiscal management of the utility. Policies should reflect the legal and regulatory requirements placed on the utility by federal, state, and local laws and regulations. The management style and local involvement of ratepayers and interested citizens may affect the nature and depth of financial-management policies. For example, separate accounting for the collection, use, and disbursement of system development fees is not only a financial best management practice but a legal requirement in some states. Another example of regulatory requirements affecting financial-management policies is the adoption of Proposition 218 in California (known as the Right to Vote on Taxes Act of 1996). With Proposition 218, the California electorate passed laws requiring that fees and charges be based on a clearly defined statement of actual needs that the utility must identify publicly and distribute to all ratepayers before consideration and adoption. Resiliency is an increasingly common concern for utilities given water supply and force majeure events. Florida recently adopted new laws to expand Senate Bill 1954 (Florida Statewide Sea Level and Flooding Rise Resilience, 2021), which established the Resilient Florida Program that was expanded, promoting the formation of resilience districts and encouraging local governments to create financial reserves specifically for resiliency projects. Finally, the acquisition and funding of new and replacement infrastructure further demands that a utility clearly identify its needs and develop sound policies related to the long- and short-term funding of those needs, always with an eye toward maximizing service life at the lowest possible cost.

The utility policies shall be set by its governing board. In developing new policies, the management and governing board may start by developing a list of the proposed operating policies that are needed by the utility. After discussions, management may elect to prepare draft policies for consideration by the public before the governing board formally adopts them. Many utilities accomplish this task during an annual planning retreat, over several different board meetings, or by use of a small committee of the governing board. Staff may want to use a separate subcommittee of the governing body or a blue ribbon committee of knowledgeable citizens to assist with policy development. At the conclusion of these evaluations, the governing body should formally adopt these policies so that the entire organization understands and implements the adopted program. As a reference point, Figure 3.1 presents an example of a resolution before the board of directors

(*text continued on page 34*)

BEFORE THE BOARD OF DIRECTORS

OF

DELTA DIABLO SANITATION DISTRICT

Re: Modifying Procedures Related to)

District Fund Accounting) RESOLUTION NO. 13/00

THE BOARD OF DIRECTORS OF THE DELTA DIABLO SANITATION DISTRICT HAS DETERMINED THAT:

WHEREAS, staff has conducted extensive research into the various operations of the District from inception in 1979 to the current time to determine the appropriate receipt and application of funds to the various operations of the District; and

WHEREAS, this research resulted in the Board establishing several funds for the receipt and expenditure of monies the District received; and

WHEREAS, the Board did, at the same time, establish fund balances in each of the funds of the District and established policy related to use of these funds; and

WHEREAS, it is appropriate to establish and modify procedures for the operation of the funds so that clear, well defined lines of responsibility are established which will protect and smooth the handling of District fund accounting; and

FIGURE 3.1 Example of Resolution Before the Board of Directors of the Delta Diablo Sanitation District in Antioch, California

WHEREAS, the Board entered into significant agreements in April 2000 for the purpose of providing Title XXII recycled water to two power plants for financial and beneficial use purposes; and

WHEREAS, the State legislature modified the bid requirement limit for new District construction projects from $25,000 to $35,000; and

WHEREAS, the Board authorized the General Manager/District Engineer to execute miscellaneous agreements for budgeted services and supplies under $100,000 annually to operate and maintain the District's Wastewater Treatment Facilities, and now the Recycled Water Facility at the July 12, 2000 meeting; and

WHEREAS, the General Manager needs the authority to settle general liability and workers' compensation claims less than $35,000 in order to meet the needs of the District's claims administrator.

NOW, THEREFORE, THE BOARD OF DIRECTORS OF THE DELTA DIABLO SANITATION DISTRICT DOES HEREBY RESOLVE AND ORDER AS FOLLOWS:

The following procedures shall be established relating to the operations of all the various programs and their related funds of the District and these procedures shall be reviewed as needed to determine their appropriateness and timeliness.

1. The General Manager shall each year prior to June 30, submit to the Board a proposed budget balancing revenues and expenditures for each of the District's program funds.
2. The Board shall consider and adopt no later than the first meeting of July the budget for that fiscal year.
3. Formal budget integration shall be employed as a management control device throughout the year.

FIGURE 3.1 *(continued)*

4. The Board shall retain the authority for the following:

   - Approval of program transactions and transfers larger than $35,000 not specifically included below.
   - Write-off of uncollectible receivables and unusable inventory.
   - Final approval of carry forward items into the new fiscal year from a previous year.
   - Establishment, combination or elimination of program funds in the District accounting system.
   - Non-recurring transfers made in compliance with special statues or ordinances which do not qualify as revenues or expenditures to the receiving or disbursing funds.
   - Approval of the District budget and any revisions thereto during the fiscal year.
   - Approval of the District audit.
   - Execute miscellaneous agreements for budgeted services and supplies exceeding $100,000 annually to operate and maintain the District's Wastewater Treatment facilities, Recycled Water facilities and Household Hazardous Waste facilities.
   - Transfer of cash or assets between operating funds, projects or budgets.
   - Disposal of fixed assets.
   - Approval of monthly financial and investment reports.
   - Approval of increases in expenditures for department budgets in the general fund and for project budgets in other funds.
   - Annual review and approval of a District investment policy.
   - Authorization of long-term obligations on behalf of the District from one or more programs or funds of the District.

5. The General Manager shall have the following authorities:

   - Preparation and implementation of the budget once adopted by the Board.
   - Reallocation of funds between line items in individual department budgets and projects that do not result in an increase in the approved budget.

FIGURE 3.1 *(continued)*

- Approval of contracts, agreements and expenditures up to $35,000 for items previously approved by the Board in the budget.
- Execution of miscellaneous agreements for budgeted services and supplies agreements and purchase orders under $100,000 annually to operate and maintain the District's Wastewater Treatment facilities, Recycled Water facilities and Household Hazardous Waste facilities.
- Payment of intermediate and progress payments on all Board approved contracts and obligations.
- Carry forward funds into new fiscal year from the previous year for budgeted, committed and/or planned expenses.
- Transfer of funds for investment purposes between funds and investment institutions based upon Board investment policy.
- Distribution of interest from investment to the funds based upon appropriate Board policy.
- Implementation of policies and procedures adopted by the Board of Directors that affect programs or fund operations.
- Settlement of General Liability and Workers' Compensation claims less than $35,000.
- Establishment of accounts and methods to properly account and manage District funds.
- Authority to sign unbudgeted agreements/contracts on behalf of the District up to $35,000 consistent with Board philosophy and direction and subject to a monthly report to the Board of all agreements/contracts signed during the previous month.
- Approval of change orders to construction projects not exceeding $2,500 and not totaling more than 10% of the approved construction contract to a maximum of $50,000. For projects in excess of $500,000, the change order limit will be established by the Board at the time of the award of the contract.
- Management of long-term debt obligations based upon the approved contracts and obligations by the Board.

- Authority to execute and/or accept easements, offers of dedication and right-of-way documents and/or drawings for the District.
- Authority to negotiate and execute contractual agreements regarding recycled water matters and issues with confirmation from the Recycling Ad Hoc Committee prior to finalization.

FIGURE 3.1 *(continued)*

PASSED AND ADOPTED on August 23, 2000 by the following vote:

AYES:	ABSTAIN:
NOES:	ABSENT:

I hereby certify that the foregoing is a true and correct copy of a resolution adopted by the Board of Directors of the Delta Diablo Sanitation District on August 23, 2000.

ATTEST: Joseph E. Canciamilla

Board Secretary

By: ______________________

cc: District Counsel

District File

FIGURE 3.1 *(continued)*

of the Delta Diablo Sanitation District in Antioch, California, that includes several policy statements developed by the District's board of directors.

After adoption of these policies, the utility can then use them to educate new employees, new governing board members, and the public. These policies should clearly define the utility's position and definitively state how the utility's financial-management system works and what internal management controls exist to ensure all parties of the utility's fiscal integrity. Policies should further allow for the clear and meaningful exposition and reporting of an integrated fiscal management position for all those interested in the utility's financial position.

As a best practice, management and the governing body should evaluate and review all policies at regular intervals to ensure that they are in compliance with new laws or regulations (Bui, 2015). Further, it may be that, as the governing board of a utility changes, the philosophy and direction of the governing board may also change, thus necessitating a change in these policies. Financial-management policies should remain flexible and current, and

(text continued on page 41)

CAPITAL IMPROVEMENT PROGRAM

POLICY STATEMENTS

GENERAL

1. It is the policy of the District to maintain a capital program which will provide collection, treatment, and disposal capacity to meet existing and future needs, while simultaneously providing for necessary renewal, replacement, and process upgrades.

2. It is the policy of the District to require developers to pay for the off-site improvements that are necessary for the developer to connect the project to the District's Subregional System.

3. It is the policy of the District to close out the CIP at the end of each fiscal year and delegate authority to the General Manager to carry forward unexpended project funds to the next fiscal year as of June 30.

4. It is the policy of the District to regularly review and adopt the necessary fees and charges that will fund the projects in the CIP.

5. It is the policy of the District to allow the General Manager/District Engineer to utilize funds from the appropriate Capital Asset Replacement program fund to replace assets that have been fully funded or whose replacement value is less than $50,000 through the Replacement Model and previously listed in a Rehabilitations and Replacements Project listing. This authority extends only to the level of funds available from the Replacement Model. The Board shall first appropriate any additional funds necessary to fully fund the replacement before any expenditure for the asset is made. Assets replaced under this policy shall be reported to the Finance Committee on a regular basis at a regularly scheduled meeting.

FIGURE 3.1 *(continued)*

6. The following items were approved under Resolution 13/00 and apply to Capital Improvement Program

a. The Board shall retain the authority for the following:

b. Approval of program transactions and transfers larger than $35,000 not specifically included below.

i. Write-off of uncollectible receivables and unusable inventory.

c. Final approval of carry forward items into the new fiscal year from a previous year.

d. Establishment, combination or elimination of program funds in the District accounting system.

i. Non-recurring transfers made in compliance with special statutes or ordinances which do not qualify as revenues or expenditures to the receiving or disbursing funds.

ii. Approval of the District budget and any revisions thereto during the fiscal year.

iii. Approval of the District audit.

iv. Execute miscellaneous agreements for budgeted services and supplies exceeding $100,000 annually to operate and maintain the District's Wastewater Treatment facilities, Recycled Water facilities and Household Hazardous Waste facilities.

v. Transfer of cash or assets between operating funds, projects or budgets.

vi. Disposal of fixed assets.

vii. Approval of monthly financial and investment reports.

viii. Approval of increases in expenditures for department budgets in the general fund and for project budgets in other funds.

ix. Annual review and approval of a District investment policy.

x. Authorization of long-term obligations on behalf of the District from one or more programs or funds of the District.

FIGURE 3.1 *(continued)*

e. The General Manager shall have the following authorities:

i. Preparation and implementation of the budget once adopted by the Board.

ii. Reallocation of funds between line items in individual department budgets and projects that do not result in an increase in the approved budget.

iii. Approval of contracts, agreements and expenditures up to $35,000 for items previously approved by the Board in the budget.

iv. Execution of miscellaneous agreements for budgeted service and supply agreements and purchase orders under $100,000 annually to operate and maintain the District's Wastewater Treatment facilities, Recycled Water facilities and Household Hazardous Waste facilities.

v. Payment of intermediate and progress payments on all Board approved contracts and obligations.

vi. Carry forward funds into new fiscal year from the previous year for budgeted, committed and/or planned expenses.

vii. Transfer of funds for investment purposes between funds and investment institutions based upon Board investment policy.

viii. Distribution of interest from investment to the funds based upon appropriate Board policy.

ix. Implementation of policies and procedures adopted by the Board of Directors that affect programs or fund operations.

x. Establishment of accounts and methods to properly account and manage District funds.

xi. Authority to sign unbudgeted agreements/contracts on behalf of the District up to $35,000 consistent with Board philosophy and direction and subject to a monthly report to the Board of all agreements/contracts signed during the previous month.

xii. Approval of change orders to construction projects not exceeding $2,500 and not totaling more than 10% of the approved construction contract to a maximum of $50,000. For projects in excess of $500,000, the change order limit will be established by the Board at the time of the award of the contract.

xiii. Management of long-term debt obligations based upon the approved contracts and obligations by the Board.

FIGURE 3.1 *(continued)*

xiv. Authority to execute and/or accept easements, offers of dedication and right-of-way documents and/or drawings for the District.

xv. Authority to negotiate and execute contractual agreements regarding recycled water matters and issues with confirmation from the Recycling Ad Hoc Committee prior to finalization.

WASTEWATER EXPANSION FUND

1. It is the policy of the District to annually review and set Capital Facility Capacity Charges for each zone to pay for the cost of necessary improvements.

2. It is the policy of the District to require each zone to pay for necessary repair and replacement of existing collection system facilities. The District will collect a Collection System Surcharge from the residents of the appropriate zone.

3. It is the policy of the District to maintain separate accounting by zone for Capital Facility Capacity Charges and repair/replacement and upgrades (surcharges).

WASTEWATER CAPITAL ASSET REPLACEMENT - FUND 089

1. It is the policy of the District to collect separate annexation fees and allocate the funds to the Capital Asset Replacement Fund.

2. A project, entitled "Unanticipated Capital Asset Replacement Project", is established for the purpose of appropriating funds for projects that are less than $10,000. The General Manager/District Engineer may approve projects up to $10,000 and will transfer funds from these appropriated funds to a separate project. A report of any fund transfers from these appropriated funds into a new project account will be made to the Board of Directors at the next Board of Directors meeting.

FIGURE 3.1 *(continued)*

The Board of Directors shall approve the use of these funds for projects greater than $10,000. The Board will authorize the transfer of funds to the new project.

3. As approved by the Board of Directors under Resolution 15/00, at the close of each fiscal year, excess profits or losses from the Wastewater General Fund will be transferred into the Wastewater Capital Asset Replacement Fund.

4. Upon retirement of the 1979 Sewer Revenue Bonds in the 03/04 fiscal year, property tax (Ad Valorem) revenue shall be split between capital asset and capital asset replacement. The split will be as follows; $250,000/Year to Capital Asset and the balance of property tax revenue to Capital Asset Replacement.

WASTEWATER CAPITAL ASSET - FUND 099

1. A project, entitled "Small District Capital Asset Project", is established for the purpose of appropriating funds for projects that are less than $10,000. The General Manager/District Engineer may approve projects up to $10,000 and will transfer funds from these appropriated funds to a separate project. A report of any fund transfers from these appropriated funds into a new project account will be made to the Board of Directors at the next Board of Directors meeting.

   The Board of Directors shall approve the use of these funds for projects greater than $10,000. The Board will authorize the transfer of funds to the new project.

2. It is the policy of the District to allocate the District's annual Ad Valorem property tax revenues to the annual payment of the 1979 Sewer Revenue Bonds. Any excess tax revenues shall be allocated to the Capital Asset Fund. After the bonds are retired, as scheduled in FY 03/04, the annual Ad Valorem property tax revenues will be allocated to the Capital Asset Fund in the amount of $250,000 and the remainder to the Capital Asset Replacement Fund.

FIGURE 3.1 *(continued)*

RECYCLED WATER EXPANSION FUND

1. All RWF Expansion shall be funded from future customers of the Recycled Water Enterprise.

RECYCLED WATER CAPITAL ASSET FUND

1. **ALL RECYCLED WATER CAPITAL ASSET EXPENDITURES SHALL BE INCLUDED IN THE DISTRICT'S ANNUAL 5-YEAR CAPITAL IMPROVEMENT PROGRAM WITH FUNDING THROUGH THE RATE STRUCTURE OR THROUGH DIRECT PAYMENTS BY LOS MEDANOS ENERGY CENTER AND THE DELTA ENERGY CENTER, ON THE TERMS AND CONDITIONS APPROVED BY THE RWF TECHNICAL ADVISORY COMMITTEE.**

2. All recycled water asset projects that involve wastewater assets will be shared on an equitable basis related to the benefits each customer base receives from the projects.

3. All assets or other items that are shared by the RWF enterprise and by the Wastewater Treatment Plant enterprise (or other enterprise of the District) shall be charged to the RWF on a proportional basis, using average cost.

RECYCLED WATER CAPITAL ASSET REPLACEMENT FUND

1. This fund shall be operated for the renewal and replacement of equipment and facilities related to the recycled water program with a useful life of 25 years or less on such terms as are approved by the RWF TAC. These facilities shall include the flow equalization basin, return water pump station and RWF share of the disinfection facility at the wastewater plant. Other equipment and facilities will be added if they directly benefit the RWF. The District initially accepted the RWF facility in April of 2002.

FIGURE 3.1 *(continued)*

2. The RWF TAC shall review and approve the useful lives of all equipment and facilities included in the funding model.

3. RWF assets with a useful life exceeding 25 years will be handled on a direct-bill basis to the RWF customers and when needed and on such terms and conditions as is approved by the RWF TAC in conformance with the Agreements for the Purchase and Return of Recycled Water.

4. After the 23rd anniversary (July 2024) of the Commercial Operations Date of LMEC and prior to the 24th anniversary (July 2025), the power plant operators and the District will conduct an assessment of the facilities to determine the need for replacement and rehabilitation and shall outline the necessary improvements and their estimated costs. The power companies shall then have until July 2026 to issue a written Notice of Intention regarding their involvement in the replacement and rehabilitation of the facilities going forward.

5. All assets or other items that are shared by the RWF enterprise and by the Wastewater Treatment Plant enterprise (or other enterprise of the District) shall be charged to the RWF on a proportional basis, using average cost.

FIGURE 3.1 *(continued)*

senior management should communicate to all concerned parties that these are dynamic and reflect the "real operations" of the utility. Management may schedule regular assessments of the utility to ensure that the utility is incorporating all policy directions in its day-to-day operations.

3.0 FINANCIAL PLANNING

Long-term financial planning is a critical element of achieving financial sustainability. Regular updates to a financial plan can minimize ratepayer shock while addressing changing capital needs; demonstrate to the rating agencies and financial markets the utility's commitment to sound financial practices; and provide a measure of reassurance of stability to employees. Utility finance is a long-term endeavor; therefore, proper long-term financial planning should be a top priority for policymakers in charge of utility systems. This section will discuss how utility decision-makers should plan for future operating budgets and multiyear capital budgets. Utilities operate in both an internally and externally dynamic environment and need to be able to adapt to future conditions with predictability.

In the context of this chapter, a long-term financial plan is defined as a financial forecast projecting at least 5 years of financial performance. There are a couple of reasons why projecting 5 or more years of financial performance is considered a best practice by the Government Finance Officers Association (GFOA) and professionals across the utility industry. Long-term financial plans compel utilities to consider long-term trends and potential risk factors so that changes can be addressed proactively. For utilities, one important benefit to the long-term financial plan is that it will facilitate a capital improvement plan (CIP) funding exercise to ensure that funding sources are available to meet capital needs. Another significant benefit is that it will provide insight to the level of future rate revenue increases.

The basis of the projection of revenues and expenses for the long-term financial plan should be based on a properly detailed annual budget or actual historical performance. When developing a long-term financial plan, a utility should consider the following issues when forecasting revenues and expenses:

- Inflation/deflation. The cost of commodities such as labor, materials, chemicals, and energy will change in future years. Commodities make up a substantial portion of a utility's operation and maintenance (O&M) budget, so incorporating changes to these costs is critical to a financial plan.
- Employee benefits. Rising health care costs and pension obligations represent another significant cost for public agencies.
- Capital reinvestment. As a utility's assets depreciate, the need to fund asset replacement to address aging infrastructure increases. There may be critical components of a utility's system that are reaching the end of their useful life in the near or distant future.
- Operational capacity. As a system grows, additional assets may be needed, and their operating costs should be incorporated to the utility's budget.
- Regulatory requirements. As the regulatory landscape evolves (e.g., effluent quality, energy usage, and carbon footprint), a utility may need to update and/or replace assets that were previously sufficient to meet regulatory burdens.
- Demand projections. As a utility's service area grows (or shrinks), the demands put on its system may change and necessitate budget adjustments.

- Frequency of revenue increases. In some areas of the country, utilities may adopt annual revenue increases that match economic indices (e.g., Consumer Price Index [https://www.bls.gov/cpi/] or Engineering News Record [https://www.enr.com/]) to minimize the revenue/expense gap and maintain financial stability. Other utilities may conduct regular financial reviews to determine a period of revenue adjustments. Regardless of the approach taken, regular revenue adjustments are a best practice of any financial-planning effort.

To project the various effects caused by the aforementioned issues, a utility should develop and maintain a financial model that details the utility's financial processes and is sufficiently flexible to reflect the ever-changing operational environment. At a minimum, a utility's financial-planning model should forecast necessary revenue adjustments needed over the planning period; calculate the debt coverage level for all outstanding debt obligations; show levels of pay-as-you-go funding, reserve funding, and transfer into/out of the operating fund; and determine the utility's annual rate revenue surplus/deficit position. More advanced financial-planning models may include the ability to run scenario ("what if") analyses to address different operating conditions, changes in the CIP and its funding, and timing of revenue adjustments.

4.0 BUDGETING

For wastewater utilities, the development of a budget must involve the evaluation of many factors that will have a direct effect on the definition of the financial resources, personnel, and supplies needed to deal with the highly regulated operations of agencies that have long-lived and capital-intensive assets. The budget is the tool that allows the utility to accomplish its required activities and set rates appropriately to support those activities. The process of budgeting at a municipal utility is one that differs from private wastewater agencies in that the process is often public and political. Many states have legal requirements that demand the development of rates and charges using a cost-of-service basis incorporating specifically defined industry standard processes. In any case, a utility's budget needs to have sufficient detail to be the foundation of properly calculated rates.

The budgeting process should allow for a complete and thorough analysis of the resources needed to meet wastewater utility requirements. It should also allow for a public process that both educates and allows for meaningful citizen input regarding utility goals and performance criteria that ultimately will translate to rates and charges. Finally, the budgeting process should align the governing body, management, and all levels of the utility staff around well-defined and clearly stated goals and objectives for utility programs and services.

Because of the nature of wastewater services and the capital-related expenditures required to operate a wastewater program, a good financial-management system delineates the "how, who, and what" of budget production. Operation and maintenance related budgets take much different forms than capital-related budgets. Typically, O&M budgets require details beyond those required for capital projects, for example, the addition of a clarifier or several miles of new pipelines. In most instances, expenditures related to O&M are relatively consistent based on the treatment of the wastes received from the service area. Capital budgets have differing components of cost, which often represent multiyear development and implementation and involve different funding sources. Preparation of separate operating and capital budgets is the typical method used by public agencies. Another significant difference is that the O&M-related budget typically appropriates the necessary funds for actual implementation, whereas the capital budget may only identify the funding and expenditure plan but may not actually appropriate funds for the full construction of all identified improvements.

Several basic decisions and guidelines must be available before initiating the budgeting process. There must be agreement about the period used for the budget (e.g., annual, biannual, or other). For many utilities, state or local requirements or debt instrument contractual provisions dictate the budget period. When selecting the budget period, a good financial-management system considers the amount and effort required of utility staff to prepare and produce the final work product. The size of this work effort may well justify the use of multiyear budgeting, especially if the utility does not experience significant changes in either revenues or expenditures from year to year.

The next step in developing a budget addresses the following questions:

- What form will the budget take?
- Will the budget include both O&M and capital?
- If the utility chooses to budget O&M and capital separately, will the utility present them together and in a generally comparable format?
- What type of information should the budget include so that people responsible for reviewing and approving the budget can clearly understand and assess the requests and program needs?
- Will the budget information tie directly to the performance goals and strategic mission of the utility? If so, how will staff know what the

strategic initiatives of the governing board and public are so that they can plan and estimate a proposed work program and decide the level of program implementation?

- What is the process for reviewing and considering the large amount of information necessary for all parties to reach agreement and consensus on the final budget?

Regulated bodies may dictate the reporting format and structure of budgets for regulated utilities. Financial management associations such as the GFOA also provide guidance on budgeting.

The utility should answer the aforementioned questions before it attempts to prepare a budget. The formal budgeting process should include a clearly defined protocol for review and approval that identifies necessary stakeholders to be involved, and a detailed schedule of the actions and activities leading to final approval should be available to all participants before the budgeting process begins. This will help keep the process on track and should allow all participants to be timely and complete in their contributions. Finally, the answer to the question of the time period covered by the budget will provide a clear and definable process for the development of any rates and charges that will come directly from the processes outlined in subsequent chapters of this manual.

Once the utility has answered the aforementioned questions, it must determine what form of budgeting to follow, that is, should the budget represent the programs of the utility; should it represent the functions of the utility; or should it reflect the specific activities necessary to accomplish only the collection, treatment, and disposal of wastewater from the service area. There is no one perfect method for preparation of a utility budget. Clearly, the purpose of the budget and the needs of the governing body related to the administration and philosophic oversight of the utility will affect decisions made while preparing the budget. In addition, the utility should consider the following questions:

- How will staff be responsible for day-to-day implementation of the budget?
- What use will staff make of the information for performance evaluations both internal and external to the organization?
- Will ratepayers or other agencies in the local service area that may serve as comparisons to the utility understand the information?
- How will management use the information to evaluate the efficiency and economics of the utility's operations and capital additions?

Several professional organizations in the industry have defined the various types and styles of budgets available to a utility. The Governmental Accounting Standards Board and the International City/County Management Association are good sources for more specific information on the types and styles of the following budgets:

- Line-item budget. A budget that identifies basic operating needs of the utility such as salaries, chemicals, supplies, benefits, utilities, and contractual and professional services and establishes spending plans for all operating units based on the needs of each unit. This is the traditional utility budget form in the industry. With this type of budget, preparers need to exercise extra care developing a line-item budget to avoid simply (and blindly) applying previous year results; rather, the budget should be based on actual needs and demands of the organization.
- Activities-based budget. A budget process that is based on specific activities of the organization or utility. This might include separate budgets for primary treatment, sludge and biosolids handling and disposal, collections system operations, or other activities defined as important to the successful understanding of an organization. This type of budget includes all costs and expenditures for the activity within budget units (e.g., salaries, utility costs, chemicals, equipment, and asset replacement). A benefit of an activity-based budget is the development of efficiency and effectiveness performance measures to help manage the financial resources of the utility.
- Zero-base budget. This method requires all expenses to be justified in each budget period, analyzing needs and costs starting from a cost base of zero. This budget method allows for prioritizing all expenditures to a final level of budgeting and allows for the selection of each set of expenditure against all others that the elected board or council must evaluate. This method avoids the setting of a budget simply based on historic precedent. When complete, a zero-base budget may have a format similar to that of a line-item or activities-based budget, but there is independent justification for each expense regardless of the results of the previous budget period.
- Capital budget (or CIP). Typically, this budget is a multiyear, long-term financial plan (5 to 10 years) for the development, studying, design, construction, and implementation of programs and projects for infrastructure and large capital investments for the support of the community. Utilities often prepare capital budgets to support a bond issue or system development fees. It is important that the utility

coordinates capital budgeting activities with the operating budget to ensure that it includes the costs of operating new facilities.

- Performance budget. This is a budget based on specific performance standards for the total unit and subunits of the budget and estimated based on anticipated performance results.

An important aspect of the budgeting process is the accurate tracking of revenues and expenditures. Utilities must consider the distinct demands placed on their systems by different customer groups, such as those requiring reclaimed water, extraterritorial services, or high-strength dischargers. The way agencies define and derive revenues for specific purposes necessitates careful decisions about revenue and expenditure tracking. To address these varied demands, utilities might segment their budget into individual units tailored to each customer group's needs.

The efficient completion of the budgeting process is critically important to the success of any organization. The budget will address the allocations of resources to divisions and departments, set the stage for both long- and short-term acquisition of capital needs, and ensure the future ability of the utility to support changes and growth in the service area. The budget should directly align expectations of the public, the governing board, and utility staff. The process should clearly explain the effects of federal, state, and local regulations on utility operations. Finally, a properly developed budget will form the basis of preparation of user charges and system development or impact fees.

5.0 COST ACCOUNTING AND FINANCIAL REPORTING

Previous sections of this chapter have addressed the issues of financial planning and budgeting. This section will address cost accounting and financial reporting that necessarily follow and must strongly support policies and budgets and that allow those who have budget oversight and responsibility know how they are doing regarding decisions made in development of the budget. As mentioned previously in the discussion of both planning and budgeting, issues and policies related to these topics will rely on the information and reporting needs in this area of an agency's operations. Federal laws and directives issued by organizations such as the Government Accounting Standards Board and Securities and Exchange Commission and state laws affecting either public or private utilities will provide much of the direction in terms of what is required of a financial accounting and reporting function. Decisions made by the governing body at the time of issue of either public or private debt may require certain reporting requirements for the utility. If a utility has received any type of federal or state grants or loans for the construction or operation of its infrastructure, then facilities may necessitate the collection and reporting of revenues and expenditures in a certain way for ease of understanding to the agencies loaning or granting funds. Private wastewater agencies may have requirements placed on them by regulating bodies. Finally, requirements in state or federal law related to the annual reporting of audit information may place certain requirements for information collected and ultimately reported for public review.

The financial and management reporting functions of an organization are some of the most complicated and diverse operations of any public or private utility. From these functions come utility audit compliance; performance assurance and accountability for the governing body and the public; justification for the rates and charges that a utility imposes; assurance of utility liquidity and financial stability over time; historical information for justification of O&M needs, including asset replacement and rehabilitation; demonstration of compliance with debt-issue requirements; and, finally, public and ratepayer creditability and trust. The level of detail and type of report varies with the audience. For example, governing boards may be interested in seeing high-level dashboards summarizing progress against key performance indicators (financial, operational, etc.). On the other hand, the finance committee may be more interested in budget reconciliation reports, whereas members of the engineering committee may wish to monitor sanitary sewer overflow occurrences and remediation activities. The references cited at the end of this chapter provide further insights to the types of financial-management reports generated typically. The financial management and reporting system must be flexible and open to the changing needs of an organization. Most accounting and financial-reporting functions are now software based. The current state of technology in this area is more software developer than customer driven and, as such, what software developers choose to provide may offer these reporting functions more than what a utility really needs. When contracting with any financial-management software provider, it is important that a utility clearly state the level of flexibility and adaptation necessary for the utility to maximize the software's usability.

The utility should establish easy and functional principles and practices for all levels within and without the organization. A utility needs flexibility and the ability to make tradeoffs in the area of accounting and financial reporting.

As a utility begins the process of defining its financial management and reporting function, it must first define the various operational centers needed for the recognition of revenues and expenditures. The definition of a *fund* is an independent fiscal and accounting entity with a self-balancing set of accounts that are segregated to conduct specific activities or attain certain objectives in accordance with special regulations, restrictions, or limitations. Typically, most governmental agencies will use a fund basis for accounting. Three main fund categories exist: governmental, proprietary, and fiduciary. These three fund categories can be broken down even further to the fund types shown in Table 3.1.

Most governmental wastewater utilities use the enterprise fund method of accounting because user fees support the majority of operational costs. Wastewater agencies may also use capital project and debt-service funds because of the nature of large capital expansion and asset needs of wastewater operations. Some utilities with large CIPs break their capital project funds into fund groups such as capital expansion, capital asset acquisition, and capital asset replacement. Creating subfunds of this nature can assist in meeting certain regulations, particularly those related to the handling of all revenues and expenses related to development, development impact fees associated with the construction and acquisition of new capacity for growth, and growth-related activities within a utility's service area. These subfunds may roll directly into the wastewater enterprise fund or they may stand alone, depending on the type of utility and the accounting philosophy of the organization. In many cases, trust indentures associated with debt issuances dictate the type of funds established.

After deciding on the fund structure, the utility must next determine the basis of accounting. *Basis for accounting* refers to the time at which revenue and expense recognition occurs and is ultimately reported in financial statements. This decision will have a direct effect on the recognition of revenues and expenditures for rate-making purposes. The two most common methods used are the accrual and cash methods. The accrual method recognizes revenue and expenses when the utility earns revenue or incurs

TABLE 3.1 Breakdown of Governmental, Proprietary, and Fiduciary Fund Categories Into Further Fund Types

Governmental	Proprietary	Fiduciary
General	Enterprise	Trust and utility
Special revenue	Internal service	
Capital project		
Debt service		
Special assessment		

expenses. For instance, in the case of user charges based on metered water use, revenue recognition occurs at the time of water consumption (when the utility provides the service), not when the utility receives the actual cash from the property owners. The cash method, on the other hand, only recognizes revenue and expenditures when the utility receives the money or the expense is paid. In the aforementioned example, revenue recognition occurs at the time the utility receives the property owner's payment.

Once the utility has determined the accounting fund structure and method of accounting, it can then develop a detailed chart of accounts that will identify how and at what detail level the utility will recognize revenues and expenses. In some states, utilities must use what is known as a "state chart of accounts" as set by statute. For regulated utilities, many regulatory bodies require the use of the National Association of Regulatory Utility Commissioners (NARUC) chart of accounts. Bui (2015) also suggests using NARUC's chart of accounts because its structure allows better support of cost-of-service rate development. The detailed chart of accounts should represent both the organizational philosophy and policies. Structurally, the chart of accounts should be of sufficient granularity to assist the agency's staff in carrying out and implementing the various programs and activities associated with utility operations; it should also provide management information required by stakeholders.

6.0 FINANCIAL RESERVES

The practice of "living paycheck to paycheck" is not a sound fiscal management approach for an organization responsible for collecting, treating, and disposing of wastewater and maintaining public health and safety. Although the concepts of establishing and maintaining financial reserves can lead to politically charged situations for public agencies, financial reserves are fundamental to a well-run and well-financed utility (American Water Works Association, 2017; Bui, 2015; Moody's Investor Services, 2014; Standard & Poor's, 2016). Utilities should clearly identify the particular use for each fund to avoid the perception that they may have collected more than the utility needs. Clearly defining the reasons for holding extra cash and directly relating those reasons to operations, emergencies, and risk management activities that may affect the utility is good public policy. Reserves for different purposes should be separate and distinct within an agency's finances.

Typically, financial-management boards define the *maintenance of reserves* as the maintenance of cash or financial capabilities to meet unknown changes in the budgets and/or previously unexpected financial needs of a utility. These needs could arise from new laws and regulations, natural disasters, operating emergencies, financial losses in earning potential from idle cash, a drop in economic conditions in the service area or state, insurance losses, litigation, revenue collection process breakdowns, or operating emergencies in the service area. These needs may vary according to the types and designations of the funds collected or expenses required, thereby necessitating different policies for varying areas or funds of an agency's operation. For example, a utility may have special requirements for funds collected to pay for new, expanded facilities that are much different than the policies for replacement of equipment or to pay employee compensation and benefit responsibilities. In some cases, utilities may restrict reserves for specific uses or purposes so that they are accessible only under predefined circumstances. In other cases, reserves may be unrestricted and available for a wide range of uses.

A well-managed and well-operated organization should take the time to identify those conditions and situations (e.g., "risks") that could have a substantial financial effect on the utility's revenues and expenditures, whether through the agency's own actions or by actions of others that can happen quickly and without significant warning. The utility should identify and determine the expected consequences of any action that would have a material effect on any of the operating revenues or expenses of the utility. In fact, criteria in being awarded a GFOA Distinguished Budget Presentation Award include explaining long-term financial plans and assumptions and strategies to address long-term concerns and issues.

Legislative actions (such as California's Proposition 218 and Proposition 13 [1978]) may establish reserve levels but, more often, reserve levels are a result of sound financial planning or requirements established as a result of revenue bond issues. Industry associations consider establishment of reserves for replacement and rehabilitation (R&R) activities an asset-management best practice. The GFOA specifically calls out maintaining appropriate levels of working capital as a best practice (GFOA, 2011).

The Water Environment Federation has identified the following possible types and uses for reserves in utilities and enterprise operations (may have different names in different regions):

- General operating reserves (including contingency reserves; reserves for claims and losses; reserves for bad debt allowances; and reserves for employee benefits, retirement, and medical funds). Utilities establish these reserves to meet general or specific operating needs. Typically, they are available for unforeseen fluctuations in expenses or revenues or to supplement rate revenues if needed to meet operating expenses. Typically, a general operating reserve is in the range of 1 to 3 months of operating costs, depending on the instability or unpredictability of revenues and expenses. For example, utilities that collect revenues on a semi-annual basis (from property taxes) may have operating reserves that are 6 to 9 months' worth of operating expenses; on the other hand, a utility that collects revenues monthly may only have 2 months of O&M in its operating reserve. Establishing these reserves or building them up to targeted levels can be done in a single year or span multiple years. Because utilities will set these reserves typically at a percentage of costs, annual additions are relatively small after the initial funding. Financial-management groups such as GFOA and rating agencies like Moody's and Standard & Poor's routinely provide guidance on best practices for operating reserve levels.

- Emergency capital reserves (including for capital construction, system replacement, line replacement, and disasters and emergencies) to cover unforeseen capital needs. The purpose of these reserves is to fund activities should a portion of the system realize an unexpected failure. Levels for these types of reserves can be set at a percentage of asset value (e.g., 1%) or equal to the replacement cost of a critical or significant portion of the facilities. Typically, these reserves are for emergency situations such as natural disasters (earthquakes, tornadoes, hurricanes, etc.) or a significant sewer line collapse.
- Rate stabilization reserves. Utilities sometimes establish these reserves in instances where there may be wide fluctuations in revenues from season to season or year to year. The reserve's purpose is to cover variations in cash flow to minimize increases/decreases to customer rates and charges. For utilities that set fixed rates for multiyear periods, these reserves can be used to accumulate surplus revenues or to address potential deficits.
- Bond covenant reserves. These reserves meet provisions of a utility's bond resolutions, covenants, or ordinances. With revenue bond issues, they often include an O&M reserve (similar to the general operating reserve discussed previously) and debt-service reserves. Typically, the debt-service reserve includes a year's worth of principal and interest. The utility may only use the funds if there are insufficient revenues to make bond payments. The proceeds of a bond issue often fund a debt-service reserve.
- Insurance reserves. Utilities may use these reserves to pay for unforeseen claims or known future expenses. Rather than purchase commercial insurance, some utilities choose to self-insure some portions of their operations. In these cases, the proceeds from such a fund can be used towards any claim, but utilities should still consult with a risk-management professional to ensure that the utility is properly protected against force majeure or other potential insurable events.

Another type of insurance reserve may relate to employee benefits such as deferred compensation or other benefit programs.

The level or need for a reserve depends on a utility's ability to collect offsetting revenues and/or manage expenses sufficiently to address the change in financial position. That is, if a utility does not have sufficient financial flexibility to meet an unforeseen condition that affects key performance measures (e.g., debt covenants and operational violations), then it needs to establish reserve funds. The type of reserve established depends on the area in which the utility has risk. For example, if a utility is trying to catch up on deferred maintenance activities, setting up an R&R reserve may provide the funds needed to consistently address infrastructure needs. A utility should determine the maximum length of time that it might have to operate without the revenue or expense and then determine the cumulative dollar value for this period of time. The utility should then consider its ability to fund the identified balances. Finally, the utility should develop a well-outlined policy statement for consideration and adoption by its governing board. This last step allows a public debate on the advisability of the policies and allows for a good understanding of the agency's identified needs by the public.

7.0 FINANCIAL MANAGEMENT AND ACCOUNTING—CONCLUSIONS

This chapter provides an overview of a well-integrated financial management and reporting system. Some examples of typical governing board policies and budgeting philosophies have been identified in addition to the general philosophies and process for good financial management and reporting. A summary of important financial management and accounting practices is as follows:

- Management and the governing body of a utility should evaluate and review all policies at regular intervals to ensure that the utility is incorporating all policy directions in its day-to-day operations;
- A utility should develop and maintain a long-term (5 years or more) financial model that details the utility's financial processes and is sufficiently flexible to reflect its ever-changing operational environment;
- A properly developed budget will address a utility's allocations of resources, set the stage for both long- and short-term acquisition of capital needs, ensure the ability of the utility to support changes and growth in the service area, and form the basis of the preparation of user charges and development or impact fees;

- A utility's chart of accounts should be of sufficient granularity to assist agency staff in carrying out and implementing the various programs and activities associated with its operations while also providing the information required by stakeholders; and
- The maintenance of financial reserves is a fundamental characteristic of a well-run and well-financed utility, and a utility's reserves should be sufficient to provide the flexibility needed to meet an unforeseen condition that affects its key performance measures.

There are many resources available to address financial and accounting matters for both municipal and private utilities. The following organizations regularly issue best practices, pronouncements, and other guidance that affect the field of wastewater budgeting, finance, and accounting:

- Government Finance Officers Association,
- Governmental Accounting Standards Board,
- Securities and Exchange Commission,
- Financial Accounting Standards Board,
- U.S. Environmental Protection Agency, and
- International City/County Manager Association.

8.0 REFERENCES

American Water Works Association. (2017). *Reserves policy white paper.* AWWA Rates & Charges Committee.

Bui, A. T. (2015). *Financial management for water utilities: Principles of finance, accounting, and management control.* American Water Works Association.

Government Finance Officers Association. (2011). *Best practice: Appropriate levels of working capital in enterprise funds (budget and CAAFR).*

Moody's Investor Services. (2014). *Rating methodology: US municipal utility revenue debt.*

Standard & Poor's. (2016). *U.S. public finance waterworks, sanitary sewer, and drainage utility systems: Rating methodology and assumptions.*

4

Financing Capital Improvements

1.0 INTRODUCTION

Wastewater utilities are capital intensive and their annual capital expenditures are what often drive the need for additional revenues beyond the more steady, year-over-year increases in operating expenses because of cost inflation. Growth in a utility's service area, aging systems, and more stringent regulatory requirements have, and will continue to place, added pressure on utilities to secure funding and financing for their ongoing capital needs. A well-conceived capital- and financial-planning process (see Figure 4.1) is critical to ensuring that the utility has adequately prepared for the future capital needs of the wastewater utility. Such planning should allow the utility to take advantage of the full array of capital funding and financing alternatives available.

One important consideration in the capital- and financial-planning process is whether and how to finance capital improvements. *Financing* refers to borrowing money with the promise to pay back the borrowed funds with interest in the future and is often used by wastewater utilities to fund significant nonrecurring capital improvements with relatively long service lives. Traditional capital-financing sources include municipal bonds, such as revenue bonds and general obligation bonds, and loans, such as bank loans and state revolving fund loans. State and federal governmental entities, such as state agencies that administer State Revolving Fund (SRF) loan programs, the U.S. Environmental Protection Agency (U.S. EPA) through the Water Infrastructure Finance and Innovation Act (WIFIA) program, and the Federal Rural Utilities Service, offer loans for qualified projects at below-market interest rates.

Another important consideration in the capital- and financial-planning process is how to fund capital improvements and support debt financing. The *funding of capital needs* refers to the source of the underlying funds that pay for capital needs. Such funding may come from a variety of sources, including user charges, special assessments, system development charges, ad valorem taxes, and grants. User charges represent the primary source of funding for wastewater system expenses and are discussed further in other chapters of this manual. As a capital-funding mechanism, user charges are widely used to fund "routine" or recurring annual capital replacements because of the relatively consistent and recurring nature of these types of capital expenditures. Paying for capital improvements in this way is often termed "paygo" or "pay-as-you-go" funding. User charges are also used to repay debt and maintain financial performance requirements and targets, such as required or targeted cash reserves and debt service coverage.

This chapter will introduce the reader to the capital planning and financing process, identify and discuss the most commonly used capital funding

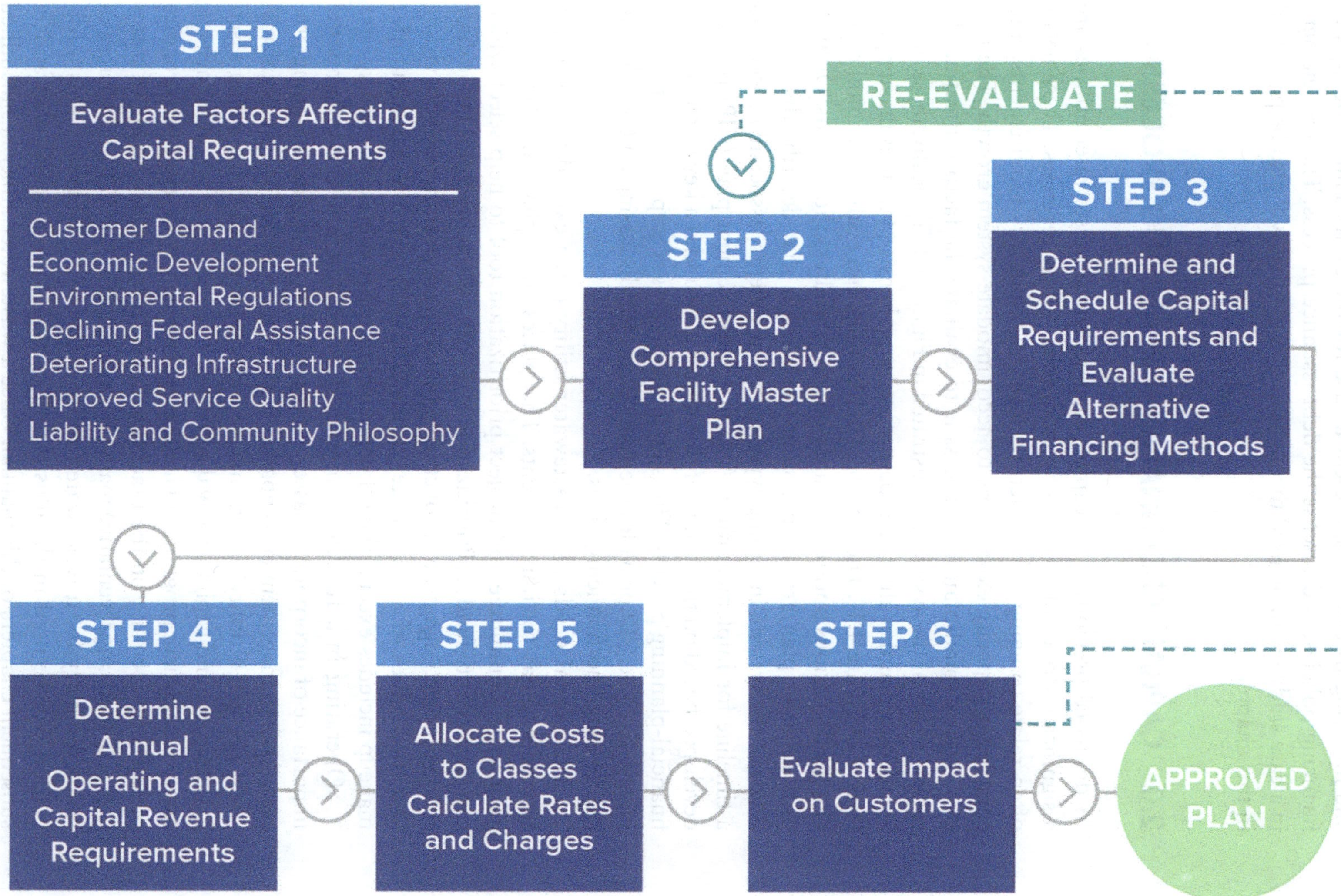

FIGURE 4.1 Capital- and Financial-Planning Process

and financing mechanisms, present a case for the establishment of capital-financing policies, and discuss the debt issuance process. The first element in a case study example used throughout this manual will also be presented in this chapter.

2.0 CAPITAL- AND FINANCIAL-PLANNING PROCESS

As discussed previously, the capital needs of many wastewater utilities are a significant driver of the need for additional utility revenues. These needs often relate to aging infrastructure, meeting regulatory requirements, and completing capacity expansions to accommodate system growth. The first step in the capital-planning process is to evaluate the factors affecting the capital needs of the system, as illustrated in Figure 4.1. Typically, these needs are identified and evaluated as part of a comprehensive facilities master plan (Step 2 of Figure 4.1). The master plan identifies and evaluates capital needs along with various options and alternatives to address those needs. Outputs of the master plan are the planning-level details that describe the capital needs of the system, including types of improvements, associated costs, and a timeline for implementation. A capital improvement plan (CIP) is often developed based on the results of the master plan and is a key element in the financial-planning process. This step in the process is Step 3 as illustrated in Figure 4.1. The CIP provides the foundation for discussing and developing the funding and financing plans.

The capital needs of a wastewater utility often exceed the available resources to pay for improvements. This places an emphasis on developing an internal committee and project prioritization tool to help select which capital needs are funded. Prioritization tools can take many forms but include scoring projects against a set of criteria that help support the mission, vision, and core values of a wastewater utility. Typically, the internal committee makeup includes executive level personnel or other subject matter experts.

Operating in a funding-constrained environment also highlights the importance of incorporating asset management to the capital-planning process. Asset management encompasses both the recordkeeping or administrative aspects of a utility's assets and provides information for ongoing asset-maintenance and capital-renewal activities. The benefits of a comprehensive asset-management program can include optimizing the life-cycle performance of assets by focusing on the practices that help to extend the life of assets (e.g., planned or preventative maintenance). An example CIP and sources of funds summary for a wastewater utility is shown in Table 4.1. This information will be used in the illustration of a capital-financing plan and in subsequent chapters of this manual. Once a wastewater utility has identified its capital improvement needs, a capital funding and financing plan can then be developed (Step 3 shown in Figure 4.1). The funding and financing plan should be developed based on the goals and objectives of the utility and the community it serves, available options and alternatives, the current financial condition of the utility, and the financial policies and covenants that it has established. These factors may dictate the use of certain funding and financing alternatives and may result in using a blend or mixture of capital funding and financing options.

A capital funding and financing plan is illustrated with the case example shown in Table 4.1. In this example, the total capital expenditures are $67 million over the 5-year study period. Of the total funding needs of $67 million, $45 million are associated with a water resource recovery facility (WRRF) expansion project. Because of the relatively long life (estimated at 30 years or more) and nonrecurring nature of this project, long-term debt financing of $43 million was identified in this plan to finance most of the WRRF project. The debt service (annual principal and interest payments) associated with the bond issue is estimated at $2.8 million per year based on a 30-year repayment term at 5% annual interest (issuance and insurance costs and a debt service reserve requirement are not included in this example). The balance of the WRRF funding requirement will be met with a $2 million grant (e.g., for preliminary design, engineering, and initial construction). Payment of the debt service is planned to be funded from system development charges. However, to the extent that capital outlays exceed the grant and bond financing amounts or if projected system development charge (SDC) revenues are not sufficient to cover the annual debt service, revenues from user charges or rates will provide the balance of necessary funding.

The capital- and financial-planning process is an iterative process, as illustrated in Figure 4.1. The evaluation of total annual revenue requirements, calculation of rates and charges, and evaluation of the effect on customers are accomplished in Steps 4 through 6, as illustrated in Figure 4.1. These steps are covered further in subsequent chapters of this manual. If, based on the CIP funding and financing plan, it is determined that the resulting customer effect is not acceptable (i.e., rates would be too high), the plan may be reevaluated. The CIP would be reexamined to determine if projects or programs could be scaled back or sequenced in a different way to reduce the effect on ratepayers. In addition to reevaluating the specific elements of the CIP, the mix of funding and financing alternatives for executing that program should also be reviewed.

The capital- and financial-planning process can involve a variety of resources that are both internal and external to the utility. Organizations at the city, county, and state level may be useful in providing demographic and related information for use in estimating the need for capital facilities.

TABLE 4.1 Capital Improvement Plan

	Year 1	Year 2	Year 3	Year 4	Year 5	All Years
Capital Outlays						
Collection mains	$1 million	$1 million	$1 million	$1 million	$1 million	$5 million
Lift stations	$1 million	$2 million	$1 million	$2 million	$2 million	$8 million
Interceptor mains	$3 million		$3 million	$1 million	$2 million	$9 million
Water resource recovery facility	$5 million	$20 million	$20 million			$45 million
Total capital outlays	$10 million	$23 million	$25 million	$4 million	$5 million	$67 million
Sources of Funds						
System development charges	$2 million	$2 million	$2 million	$2 million	$2 million	$10 million
Grants	$2 million					$2 million
Bond proceeds*	$3 million	$20 million	$20 million			$43 million
User charges—rates	$3 million	$1 million	$3 million	$2 million	$3 million	$12 million
Total funding sources	$10 million	$23 million	$25 million	$4 million	$5 million	$67 million

**Note.* $43 million bond issued in Year 1.

It may be advantageous to create a citizen committee to assist the utility in evaluating CIP alternatives, prioritizing projects, and evaluating alternative financing tools. The use of engineering resources (internal and external) is important for the development of detailed capital projects, schedules, and cost estimates. Similarly, the use of both internal and external financial and legal resources or specialists, including a registered municipal advisor, is important to examine the range of financing strategies and resources available to develop a funding and financing program and assess the plan's overall effect on utility rates and customers.

3.0 ESTABLISHING CAPITAL-FINANCING POLICIES

Capital-financing policies identify the financing alternatives available to utilities and can specify the rules and limitations associated with capital financing to protect the utility's financial condition and manage financial risk. Capital-financing policies may include the following:

- Guidelines on when internal funding and capital financing should be considered;
- Identification of financing alternatives available for consideration;
- Rules associated with the applicability of financing alternatives to various types of capital projects;
- A limit on the total debt of the utility;
- Specification of minimum financial targets and ratios;
- A limit on the use of variable-rate debt obligations, swaps, and other derivatives;
- Specification of minimum or maximum debt amortization terms; for example, a policy might specify that the debt amortization period should not exceed the average service lives of the assets being financed;
- Limitations on the use of bond proceeds for operational purposes;
- Requirements regarding the process of issuing municipal bonds, including the procurement of financial advisors and underwriters; and
- Requirements to explore opportunities to secure grants and low-cost financing alternatives.

A capital-financing policy serves to identify when internal versus external financing sources would be used. For example, a wastewater system has identified a pipeline replacement financing need of $1 million per year for the foreseeable future. Rather than issue bonds or secure loans every year or every other year, it may be more appropriate to fund this program with user charges. Alternatively, the construction of a new or expanded WRRF is an undertaking that will likely have a service life in excess of 20 years. For most systems, this would represent a significant, infrequent capital project for which bonds would be issued or loans secured. A capital-financing policy may provide guidance that would support a pay-as-you-go funding decision for the pipeline replacement program and a debt-financing decision for the WRRF. The established capital-financing policy is often the result of extensive discussion and analysis to reflect the fiscal goals and objectives of the utility. The capital-financing policy, while subject to revision, generally survives changes in city councils, utility boards, and so on.

For a wastewater system that is a component unit of or an agency within a municipality, existing citywide capital-financing policies may have direct applicability and should be considered in the development of a utility-specific set of financing policies. In the development of a capital-financing policy for a wastewater utility, consideration should also be given to longer-term capital improvement needs of the utility and the existing authority to access and use various capital-financing alternatives. Table 4.2 lists example capital-financing policies that a utility might consider.

Policies may be based on expenditure amounts (e.g., more than $1 million) and service-life thresholds (more than 20 years) for determining whether projects should be funded from internal or current revenues or financed with long-term financing vehicles, such as bonds and loans. Although there are certain general guidelines in establishing such a policy, care should be taken to specifically evaluate the CIP needs and legal authority for various financing instruments and tailor a program to the specific system. However, as a general rule, programs that should be considered for debt financing would include projects having an extended service life and a project cost above a certain threshold and projects that are undertaken infrequently. For example, a WRRF expansion with a service life in excess of 30 years and a cost of $25 million might be an appropriate candidate for the use of long-term debt financing, such as bonds. Along with the segregation of projects between cash funding and debt financing, the wastewater utility should also conduct a thorough evaluation of the eligibility of each project for any state and federal grant monies.

4.0 SOURCES OF CAPITAL FUNDING AND FINANCING

Capital funds can originate from sources that are internal or external to the utility (see Table 4.3). For example, revenues or accumulated cash reserves

TABLE 4.2 Sample Capital-Financing Policy Statements

Internal	External
Whenever possible, capital projects will be financed using self-supporting revenue bonds. Revenue bonds ensure the greatest degree of equity because those who benefit from a project and those who pay for a project are most closely matched.	Noncapital furnishings and supplies will not be financed from bond proceeds.
Bonds cannot be issued for a longer maturity schedule than a conservative estimate of the service life of the asset to be financed.	Seek guidance from a municipal advisor who can evaluate the appropriate method of sale based on the specifics of the details of the transaction.
General obligation debt will not be used for enterprise activities.	The utility will minimize the use of nonrecurring revenue sources to fund programs incurring ongoing costs.
Long-term borrowing will be confined to capital improvements.	The utility should combine a pay-as-you-go strategy with long-term financing to keep the debt burden sufficiently low to merit higher bond ratings and to provide sufficient available debt capacity in an emergency.
The utility shall seek to maintain and, if possible, improve its current bond rating so that borrowing costs are minimized and access to credit is preserved.	The utility limits debt to projects that cannot be reasonably funded in a single year and to terms that are consistent with the service life of the project being undertaken.
Short-term borrowing will not be used to finance operating needs except in the case of an extreme financial emergency beyond the utility's control or reasonable ability to forecast.	The utility seeks the least costly financing available. All debt commitments are reviewed by the treasurer, who looks for opportunities to combine issues or for alternative methods that will achieve the lowest possible interest rates and other borrowing costs.
Bond proceeds should be limited to financing the costs of planning, design, land acquisition, buildings, permanent structures, attached fixtures or equipment, and debt issuance costs.	

TABLE 4.3 Capital-Financing Sources

Internal	External
Current revenues (i.e., rates or user charges)	General obligation bonds
Cash reserves or equity	Revenue bonds
Special assessments	Green bonds
Ad valorem taxes	Special assessment bonds
System development charges	Commercial paper
Equity	Government loans and grants
Cost reductions	Public–private partnerships
Depreciation	Lease arrangements
Intramunicipal loans	

generated from wastewater rates, fees, and charges are internal funding sources, whereas cash proceeds from the issuance of debt, as well as grant funds, are external funding sources. This subsection provides a summary of various types of capital funding and financing alternatives available to wastewater utilities. A more detailed discussion of funding and financing alternatives can be found in other industry publications (e.g., *Water Utility Capital Financing, Manual of Water Supply Practices M29* [American Water Works Association, 2014]).

5.0 INTERNAL FINANCING SOURCES

5.1 User Charges

User charges are a source of funding for not only operation and maintenance (O&M) expenses of a wastewater system but also a portion of the system's capital improvements. The term *user charge* means a charge that may include any and all costs associated with the provision of wastewater service, including cash-funded capital and debt service.

This definition is more expansive than that defined by U.S. EPA, which, at a minimum, includes O&M and replacement costs. U.S. EPA purposefully excluded any direction about how the utility should fund and recover capital costs. U.S. EPA is primarily interested in the utility's ability to fully operate and maintain the system and pay for system replacement costs during the associated facilities' estimated service lives. Typically, the amount that a wastewater customer is charged is based on some measure of water use and estimated discharge to the sewer system multiplied by a commodity rate (fixed charges may also be assessed). Billed on a periodic basis (monthly, bimonthly, quarterly, semiannually, or annually), user charges can provide a relatively reliable revenue source for funding O&M expenses, debt service payments, and capital project costs.

User charges can be established to generate sufficient revenues to fund capital-improvement projects on a pay-as-you-go basis. For example, user charges may be set at a level to generate sufficient revenues to fund an annual capital program to repair or replace aging collection system pipes. Rate revenue as a capital-funding source is particularly useful for annual or recurring programs because this promotes intergenerational equity between current and future ratepayers. User charge revenues can also be used to establish reserves for funding future capital needs or for specific capital improvements (e.g., a pipeline replacement program).

As shown in Table 4.1, the amount of capital to be funded from user charges can vary from year to year based on the capital program and the availability of other funding sources. One of the challenges associated with the financial-planning process is the variable nature of revenue requirements from year to year and the desire by utilities to "smooth out" the rate-related portion of the capital funding needed and prevent spikes in wastewater rates.

5.2 Special Assessments

If wastewater infrastructure is constructed for a specific purpose or for the benefit of a defined area, special assessments may be a revenue source that is available to fund the capital infrastructure within the assessment area or district. These special improvement districts are used by wastewater agencies to install or replace local sewer collection facilities. The cost of the local facilities can be recovered by a special assessment to each property owner based on square footage of the property, adjacent or "front" footage, and other methods. In many instances, approval from the property owners is required before establishing an improvement district and assessing costs.

5.3 Ad Valorem Taxes

Ad valorem taxes are used as a capital funding source for wastewater systems in some areas of the country. The use of ad valorem taxes to fund wastewater capital improvements has greatly diminished in recent years in states like California, where legislation requires a vote to pass increases to existing ad valorem taxes or to institute new tax levies. The tax is generally levied based on the value of real estate, personal property, or both. As such, ad valorem taxes may be considered a more secure and stable funding source than user charges in areas of stable property values. The wastewater agency must have the taxing authority to use this funding method. As such, it is more applicable to wastewater utilities owned and operated by city and county governments than to authorities and special districts. Under this funding alternative, unless the ad valorem tax is specifically "pledged" to the wastewater utility, this funding source may be diverted for other general government purposes and be less reliable as a funding source than other alternatives, such as dedicated user charges.

The appeal of the ad valorem tax methodology for funding capital improvements lies in its relative simplicity in both assessment and administration. Another advantage is that the tax is also assessed to undeveloped or unsewered properties in the service area. This is a means of receiving a financial contribution from properties that may connect to the system in the future and derive a benefit from existing WRRFs.

One of the most significant disadvantages of using ad valorem tax revenues to fund wastewater infrastructure is the lack of equity associated with this method of assessment because the relationship between assessed value and the property owner's usage or benefit of the wastewater system is relatively weak. That is, the tax is assessed based on property value; a measure may not relate to wastewater flows or service demands as well as a user charge approach. Additionally, certain types of properties are accorded tax-exempt status and, therefore, would be excluded from the payment of ad valorem taxes (e.g., government-owned properties and, in some cases, property owned by not-for-profit and religious organizations); these properties could contribute a significant amount of wastewater to system but may not be responsible for sharing in the costs.

5.4 System Development Charges

System development charges, also known as *facility investment fees*, *tap fees*, *impact fees*, and *capacity fees*, are one-time payments collected typically from a developer or property owner at the time a building permit or certificate of occupancy is issued for new or expanded connections or service. The system development charge that is paid is often designed to recover a portion of the system capacity cost to serve the new wastewater system customer. Capital improvement costs associated with the expansion of the system are often paid for by new customers or recouped by the wastewater utility as the customer connects to the system and pays the SDC. As such, these expansion- or growth-related capital improvements may not need to be funded with wastewater user charges.

In a growing community, SDCs can represent a significant capital funding source, which can limit the need to fund growth-related capital projects through user charges. However, because SDC revenues are dependent on the amount of new development, this revenue source can be subject to a high degree of uncertainty and fluctuation from year to year. Furthermore, the utility may not have the luxury of waiting until a significant amount of SDC revenue has been accumulated to construct needed facilities (e.g., a WRRF expansion). As such, it may be necessary to use other forms of funding (e.g., user charges) to support the payment of the capital infrastructure. If the capital infrastructure for the expansion of the system is financed with long-term debt, a combination of SDC revenues and user charges may be needed to service the debt until sufficient SDC revenues are generated to repay the debt.

Chapter 10 presents a discussion of the underlying philosophy of SDCs and legal requirements and alternative calculation methodologies.

5.5 Equity

The financing of capital improvements via equity funds generally represents contributions made by the owners of the wastewater system. In a publicly owned wastewater system, "equity" contributions would take the form of accumulated cash generated via the assessment of user charges. In this context, equity is no different than user charges. However, for an investor-owned utility, equity funds may take the form of common and preferred stock issued by the company or direct equity contributions. In terms of priority, dividend payments on equity capital are generally subordinate to the payment of principal and interest on long-term debt. Accordingly, typically, equity funds are more expensive and require a return greater than the cost of debt financing (see the Section 8.0 "Financing Considerations for Investor-Owned Utilities" later in this chapter).

5.6 Cost Reductions

The review of business and operational practices of a wastewater utility may uncover ways to reduce O&M or capital costs and, to the extent that a wastewater system is able to reduce its costs through greater efficiencies, previously committed rate revenue may be able to be redirected for use in funding or supporting the financing of capital improvements. Cost reductions may be identified through the execution of a total quality improvement program, Lean Six Sigma program, value engineering, or other similar management and operation assessments. The ability of wastewater utilities to reduce energy costs by capturing and generating energy from biosolids is an example of how cost reductions could be used to offset the cost of capital improvements.

Additionally, in recent years, there has been significant interest in alternative project delivery methods such as design–build, design–build–operate, progressive design–build (PDB), Construction Manager At Risk (CMAR), and public–private partnerships (P3s), which may be able to lower the total life-cycle cost of an asset as compared to more conventional procurement methods. These cost savings, if realized, could provide additional capacity for the wastewater utility to fund or finance additional capital infrastructure. As such, cost reductions are often viewed as a potential funding alternative for capital projects.

5.7 Depreciation

Depreciation is a noncash expense that reflects the use of an asset over its life. Typically, depreciation is not included under the cash needs approach of defining revenue requirements (see Chapter 5 for a further discussion of the cash-needs approach). However, depreciation expense may be used to roughly approximate the minimum amount of needed repair and replacement of an existing wastewater system to maintain the condition of the system. Therefore, it is not unusual for wastewater utilities to include an amount for repair and replacement of infrastructure that is at least equal to annual depreciation in the determination of annual revenue requirements. This practice creates a revenue stream that can support the funding of system capital improvements or the contribution to a capital replacement reserve if the actual annual capital spending is less than the revenues generated that are dedicated for this purpose. Furthermore, wastewater user charges that are set to meet U.S. EPA user charge regulations include O&M and replacement costs. The replacement component of these costs can be roughly estimated by the depreciation amount.

5.8 Intramunicipal Loans

Some wastewater utilities are part of a city or county government that operates multiple utilities, such as water, wastewater, stormwater, and solid waste utilities. In certain circumstances, other utilities that are part of the municipality could make a loan to a wastewater utility to temporarily help the utility pay for a capital project or allow the utility sufficient time to phase in wastewater rate increases over time to improve its financial condition. Typically, this funding option is limited in the amount and frequency of the borrowing, depending on the financial and cash position of the other municipal utilities and if another utility of the municipality has excess or idle cash reserves. There is often a loan agreement established between the two utilities specifying the loan repayment period, interest rate, and other terms of the loan. Some municipalities may have fiscal policies or requirements that prevent the transfer of funds from one utility to another, thereby making this alternative prohibitive.

6.0 EXTERNAL FINANCING SOURCES

6.1 Public Financing

Financing capital improvements through the use of a debt can take many forms, but the most common forms are loans from both public and private agencies and municipal bonds sold to investors in the municipal securities market. One of the advantages of debt financing is the ability to spread the payment of a new facility or project (e.g., a new or expanded WRRF, as illustrated in the example in Figure 4.2) over a long-term period, often 20 to 30 years. The repayment of the debt occurs over time by inclusion of debt service payments in wastewater rates, which allows the utility to recover the cost of the project from customers as the asset or facility is used. Projects such as WRRFs often have a service life of 30 years or more and represent projects that occur with relative infrequency compared to annual pipeline projects or periodic lift station upgrades and replacements. Debt financing allows for a reasonable assignment of costs across generations (i.e., customers that benefit from or use the asset over its life cycle pay for the cost of the project). In addition to improving equity between generations of ratepayers, debt financing may allow for the construction of assets sooner than would occur if all of the cash had to be accumulated before beginning construction.

Careful financial planning is important to ensure that sufficient revenues are generated each year to provide for the annual debt service or repayment of the debt and the establishment of any required reserves. Revenues must also be sufficient to meet other bond covenants associated with the debt, if applicable, such as meeting debt service coverage ratio requirements. Typically, *debt service coverage* is defined as the amount of operating revenues, after paying O&M expenses, that are available to pay debt service. It is common for a bond covenant to have a requirement to generate sufficient revenues to meet a debt service coverage ratio of 1.2 times or more. Decision-makers should understand these annual requirements and plan accordingly.

6.2 Public Financing Alternatives

For government-owned wastewater utilities, the primary source of capital financing has been, and is likely to continue to be, tax-exempt municipal bonds. Bonds can take many forms and, in this section, the advantages and disadvantages of the various alternatives will be discussed. In addition, other

NEW ISSUE – BOOK-ENTRY ONLY

Insured Ratings: Fitch: "AAA"
Moody's: "Aaa"
Standard & Poor's: "AAA"
Underlying Ratings: Fitch: "AA"
Moody's: "Aa3"
Standard & Poor's: "AA"
See "RATINGS"

In the opinions of Co-Bond Counsel, interest on the Bonds is excluded from gross income for federal income tax purposes, is exempt from Colorado income tax and is not an item of preference for purposes of computing the individual or corporate alternative minimum taxes to the extent, upon the conditions and subject to the limitations stated under "TAX EXEMPTION."

$30,700,000
CITY AND COUNTY OF DENVER, COLORADO,
for and on behalf of the Wastewater Management Division of its Department of Public Works, Wastewater Revenue Bonds Series 2002

Dated: April 1, 2002 **Due: November 1, as shown below**

The principal of and final installment of interest on the Bonds are payable upon presentation and surrender thereof, and all other interest (due November 1, 2002, and each May 1 and November 1 thereafter) is payable by, the Manager of Revenue, ex officio Treasurer, of the City, as paying agent, by check, draft or wire transfer to the registered owners of the Bonds. The Bonds are issuable in fully registered form and are initially to be registered in the name of Cede & Co., as nominee for The Depository Trust Company, New York, New York, as securities depository for the Bonds, to which payments of principal of and interest on the Bonds are to be made. Purchases by beneficial owners are to be made in book-entry form only in the principal amount of $5,000 or any integral multiple thereof. Beneficial owners are not to receive certificates evidencing their interests in the Bonds. See "THE BONDS – Book Entry System."

The scheduled payment of principal of and interest on the Bonds when due will be guaranteed under an insurance policy to be issued concurrently with the delivery of the Bonds by Financial Guaranty Insurance Company.

Financial Guaranty Insurance Company

FGIC is a registered service mark used by Financial Guaranty Insurance Company, a private company not affiliated with any U.S. Government agency.

MATURITY SCHEDULE

Year	Principal Amount	Interest Rate	Yield[1]	Year	Principal Amount	Interest Rate	Yield[1]
2003	$ 985,000	3.500%	2.350%	2013	$1,505,000[2]	5.000%	4.650%
2004	1,015,000	3.500	3.000	2014	1,580,000[2]	5.250	4.800
2005	1,045,000	3.500	3.410	2015	1,660,000[2]	5.000	4.900
2006	1,090,000	4.000	3.700	2016	1,740,000[2]	5.250	5.000
2007	1,135,000	4.000	4.000	2017	1,830,000[2]	5.250	5.080
2008	1,180,000	4.250	4.140	2018	1,920,000[2]	5.250	5.120
2009	1,235,000	4.250	4.280	2019	2,015,000[2]	5.500	5.140
2010	1,300,000	4.500	4.380	2020	2,115,000	5.000	5.200
2011	1,365,000	4.500	4.450	2021	2,220,000	5.125	5.230
2012	1,430,000	4.750	4.540	2022	2,335,000	5.125	5.250

(1) This information is provided by the Underwriter.
(2) Priced to the first optional redemption date.

(Plus accrued interest from April 1, 2002)

The Bonds maturing in the year 2013 and thereafter are subject to optional redemption prior to their respective maturity dates as described under "THE BONDS – Redemption."

The Bonds are issued for the purpose of financing storm drainage facilities for the City. See "USE OF PROCEEDS – The Project."

The Bonds are special and limited obligations of the City, for and on behalf of the Wastewater Management Division of its Department of Public Works, payable solely from and secured by a first lien (but not necessarily an exclusive first lien) upon the net revenues derived by the City from the operation of its storm drainage and sanitary sewerage facilities. The Bonds are also payable, under certain circumstances, from a debt service reserve account. **The Bonds are not general obligations of the City. The Bonds are not payable in whole or in part from the proceeds of general property taxes nor is the full faith and credit of the City pledged to pay the Bonds.** See "SECURITY FOR THE BONDS – Pledge and Flow of Funds."

This Cover Page contains certain information for quick reference only. It is *not* a summary of this issue. Investors must read this Official Statement in its entirety to obtain information essential to making an informed investment decision.

The Bonds are offered when, as and if issued, subject to approval of legality by Ballard Spahr Andrews & Ingersoll, LLP and Bookhardt & O'Toole, Co-Bond Counsel, and certain other conditions. It is expected that the Bonds will be available for deposit with The Depository Trust Company and delivery in New York, New York, on or about April 24, 2002.

Stifel, Nicolaus & Company, Incorporated
Hanifen Imhoff Division

The date of this Official Statement is April 9, 2002

FIGURE 4.2 Example of Official Statement

public financing alternatives are discussed in this section. These alternatives include the following:

- General obligation bonds,
- Revenue bonds,
- Green bonds,
- Special assessment bonds,
- Commercial paper,
- Certifications of participation, and
- Government loans and grants.

6.2.1 General Obligation Bonds

A general obligation bond is a bond that is backed by the full faith and credit and the taxing authority of the municipality. General obligation bonds are most commonly backed by ad valorem taxes, which are based on property values. As such, general obligation bonds are one of the lowest cost alternatives available to finance utility infrastructure. Historically, general obligation debt has had widespread use by local governments and their utilities because of its low borrowing cost and ease of issuance (relative to other alternatives). This financing alternative is available to wastewater utilities that are part of a city or county government because they possess a taxing authority. This alternative is generally not applicable to wastewater authorities or special districts where the taxing authority does not exist. The interest paid to investors in general obligation bonds of a municipality is generally tax exempt if the proceeds of the general obligation bonds are used for public purposes. The interest payable to the bond holder is exempt from federal income taxes and, in many instances, state taxes.

One disadvantage of general obligation bonds relates to obtaining the authority for the issuance of the debt. In some states, voter approval is required to issue general obligation bonds; however, general obligation bonds used to fund utility assets may not require voter approval. Another disadvantage is that the debt may have an effect on the municipality's overall debt limitation and may limit the municipality from issuing general obligation debt for other nonuser-fee-based services, such as parks and recreation or fire and police.

6.2.2 Revenue Bonds

As the name implies, revenue bonds are bonds that finance projects in which the principal and interest payments are paid from the revenue generated by the project; in this case, revenue bonds are secured by revenues of the wastewater utility, not by the full faith and credit of the taxing authority of the issuing entity. Revenue bonds are available to wastewater utilities that are not part of a city or county government and do not have the taxing authority to issue general obligation bonds. Typically, revenue bonds do not require voter approval and are often not covered under a municipality's debt limit.

Wastewater revenue bonds are secured by the revenues generated from the wastewater system or project. As a result, the marketability of these bonds is dependent on the project being financed, the revenue-generating ability of the utility, and the nature and reputation of the issuer. Therefore, in general, revenue bonds may carry a higher interest cost than general obligation bonds.

Typically, revenue bonds also come with additional requirements that are not associated with general obligation bonds to provide assurance to investors that the revenues of the system can repay the debt. These provisions or covenants may require that the entity establish user charges that are sufficient to achieve a certain debt service coverage ratio (a requirement to achieve a debt service coverage ratio of 1.2 or higher is common). In general, the higher the debt service coverage requirement, the greater assurance to bond holders that revenues will be sufficient to pay the debt service.

Bond covenants may also contain a variety of other assurances to lenders that repayment will occur. These may relate to the establishment of debt service reserves and O&M and capital replacement reserves (see Chapter 3 for more information on reserves). Limitations may exist with regard to the issuance of new debt that may have a claim on revenues equal to or on parity with existing debt. For example, a parity or "additional bonds" test may have to be met before additional bonds can be issued. Typically such a test would involve a determination that the revenues of the utility for the most recently completed year were sufficient to repay the maximum annual debt service associated with all existing debt and the anticipated or new issue. Often, the bond covenants will allow for adjustments to the utility's revenues for partial year rate increases or other material changes that would affect the test.

Even in instances where the wastewater utility has been set up as a separate enterprise fund, revenue bonds may be secondarily secured by the taxing authority of the entity. In these instances, the debt is referred to as a "double barrel" bond, in that it has both a primary (utility revenues) and secondary (taxes) pledge.

6.2.3 Green Bonds

Green bonds are bonds that are specifically marketed as providing environmental or climate benefits and where the use of the bond proceeds are dedicated to projects intended to achieve these benefits. The difference between a green bond and a regular bond is that the issuer specifies or certifies that it will use the money to pay for projects with these benefits. Typically, one of the key benefits of green bonds is that they can provide the issuer with broader access to investors and greater demand for bonds than nongreen bonds because there is a significant and growing demand by investors for environmentally sustainable projects. Because of this growing investor demand, the issuer may be able to receive marginally favorable interest rates over nongreen bonds. Issuing green bonds can also help the wastewater utility highlight that they are a green-focused utility, which can lead to positive publicity for the utility. One of the drawbacks of green bonds is the cost of ongoing tracking, monitoring, and reporting to demonstrate that the bond proceeds were, in fact, used to finance green projects. Another drawback to green bonds is that the rules, regulations, and standards around what defines a green bond are still evolving, which has resulted in accountability concerns by some investors.

6.2.4 Special Assessment Bonds

With special assessment bonds, the proceeds are used to pay for capital improvements or capital projects that will benefit a defined area. In this instance, the bonds are payable only from the receipt of special benefit assessments. For example, a special assessment area may be defined to include a specific neighborhood for which individual septic tanks are being replaced by a wastewater collection and conveyance system. The individual accounts or properties are the specific beneficiaries of the project and special assessment debt could be used to ensure that only those properties benefitting are assigned the costs of these improvements.

A special assessment bond is backed by ad valorem taxes that are assessed on those who directly benefit from the facilities. Relative to the overall debt-funding requirements of the entity or community, special assessment bonds are of a smaller issue amount typically. The annual debt service is paid or secured through the annual property tax assessment for each individual account or property. Typically, assessment amounts are based on the value of each property (in similar fashion to how property taxes are determined) or on some other measure of the benefit accruing to the property (e.g., front footage or fixture unit count for each property in the assessment area or district).

6.2.5 Commercial Paper

Commercial paper is a short-term, unsecured borrowing that is used as a financing "bridge" until long-term financing can be obtained at an attractive cost. As used by a local wastewater utility or city, such borrowings will be tax exempt typically and secured by other funds such as a letter of credit.

This form of borrowing is prevalent for many large municipal utility providers because of its short-term nature and tax-exempt status and serves as an attractive means of funding capital facilities until long-term financing can be arranged. Typically, a key benefit of commercial paper is its extremely low interest rates. However, commercial paper is generally suitable only for larger issuers because of its relatively high startup and ongoing operating costs. A commercial paper program can also save resources by avoiding or delaying the development of an official statement that includes financial and engineering feasibility studies—studies that are required typically when issuing revenue bonds.

6.2.6 Government Loans and Grants

As a form of capital project funding, a grant represents funds provided to a wastewater system without a repayment requirement. Going back to the Clean Water Act of 1972, a variety of state and federal programs have existed and still exist today (primarily administered at the state level, but with federally funded resources) to assist wastewater utilities in financing applicable capital improvements. Typically, most of these programs provide capital funding assistance in the form of low interest loans.

The Infrastructure Investment and Jobs Act (IIJA) was passed in 2021 and provides $11.7 billion in funding to the Clean Water State Revolving Fund, of which 49% of the funding will be in the form of grants or principal forgiveness, as well as $1 billion in grants or principal forgiveness for emerging contaminants. The IIJA represents the largest increase in wastewater infrastructure funding in decades and is a start to addressing the lack of available federal funding that has steadily decreased since the 1970s. There are also a variety of programs administered at the state level that can provide grant funds (and, in many cases, interest-subsidized loans) for publicly owned wastewater systems (e.g., state revolving funds, U.S. Department of Housing and Urban Development, and U.S. Department of Agriculture Rural Housing Service).

The specifics of any grant or loan program can differ significantly from state to state in terms of the application and award process. In general, these programs consider the type of project (system expansion versus renewal and replacement); the financing capability of the applicant; and, in many cases, the economic status of the service area. This last criterion is often intended to define or determine economically distressed areas and the effect that traditional financing mechanisms would have on the affordability of wastewater service in the community. Funding (grants and loans) awards may be based on the amount needed to maintain the affordability of the service; however, some programs may require the utility to cover a portion of the project costs with its own funds, in tandem with the grant or loan proceeds, to secure such funding. Consideration of these and other criteria by state and federal agencies that have responsibility for the administration of these programs is one step in prioritizing funding applications and awards.

In addition, at the federal level, the WIFIA program, passed in 2014, now provides direct loans and loan guarantees for water and wastewater infrastructure projects. The program is administered directly by U.S. EPA and is separate from SRF programs but often works in coordination with these programs to provide funding for qualifying projects. Additionally, WIFIA funding is often used alongside many other financing alternatives as a source of project financing.

According to program guidelines, WIFIA funds can be used to fund up to 49% of the costs for a qualifying project. The minimum project size for large communities is $20 million, whereas the minimum project size for small (population of 25,000 or less) communities is $5 million. Some examples of eligible project costs include certain development phase or preconstruction activities; construction, reconstruction, rehabilitation, and replacement activities; the acquisition of, or an interest in, real property; and financing-related costs, including capitalized interest, required reserves, issuance expenses, and other carrying costs incurred during construction.

To apply for funding under the WIFIA program, applicants are required to demonstrate the project's eligibility, feasibility, creditworthiness, and alignment with U.S. EPA's policy priorities. Projects are selected based on a predefined set of criteria, some of which include the significance or economic and health benefits to be realized as a result of the project, the extent to which the project protects against severe weather events, the extent to which the project relies on additional public and private financing alternatives, and the degree to which the project serves to benefit communities facing economic hardship.

6.3 Private Financing

Private financing can take many forms, but the most common forms are bank loans and private debt and equity in connection with P3s. These forms are discussed in the following subsections.

6.3.1 Private Placements

Private placements include bank loans and direct purchase bonds. A bank loan can occur in two different ways. If a bank buys the bonds of a wastewater utility, it is called a *direct purchase*. If the bank loans money to the utility, it is called a *direct loan*. Direct lending through private placements is often considered if the issuer's credit is deemed to be less than investment grade.

Typically, the process of executing a loan with a bank is more streamlined than issuing traditional bonds, reducing the time required to receive the financing. In addition, there are fewer costs of issuance and fewer ongoing compliance requirements than issuing more traditional municipal bonds. Bank loans can also be structured in a more flexible way than traditional municipal bond issues. However, private placements could result in higher effective interest costs because the debt often has less liquidity than more traditional municipal bonds. In addition, these loans may also have covenants that contain acceleration clauses, requiring the loan to be repaid in full if the issuer defaults on the loan.

6.3.2 Public–Private Partnerships

Public–private partnerships are arrangements between a public entity and a private company in which the resources, risks, and rewards are combined to provide greater efficiency, better access to capital, and improved compliance with government regulations. A P3 arrangement is long term, often lasting 10 to 20 years or more. It is commonly accepted that private financing is a key distinguishing feature of a P3 arrangement (Water Research Foundation, 2016). Under a P3 arrangement, the private concessionaire designs, constructs, finances, operates, and maintains the asset and either receives the rights to future revenue streams from user charges or receives availability payments from the utility. The advantages of a P3 arrangement include allowing for risks to be shared between the public and private sectors, enabling the capital project to be implemented on an accelerated timeline, providing incentives to the private sector to innovate and reduce the total life-cycle cost of the project, and providing the utility with certainty regarding the total cost of the project (and often the future user charges to support it). One disadvantage of P3s is the cost of private financing, which can be significantly higher than the cost of municipal, tax-exempt debt. Therefore, the efficiency savings often achieved with a P3 arrangement need to be sufficient to offset the higher cost of financing for these arrangements to be economically viable. Furthermore, typically, the high upfront cost of these arrangements makes P3s suitable only for large infrastructure projects (see Chapter 14 for additional information on P3 arrangements).

6.4 Contributions

Some components of a wastewater system needed for new development may be financed and constructed by developers and then contributed to the wastewater utility. For example, the installation of the local (on-site) collection system infrastructure for a new residential subdivision is often the responsibility of the developer. After construction and subsequent inspection by the utility, these facilities are then dedicated or turned over to the utility by the developer. In some instances, a similar practice may be used for the financing of off-site facilities. Developer dedications or contributions are a common means of funding capital improvement projects for wastewater systems when the developer wants to extend service to a property where utility-planned extensions do not exist in the near term.

In development or annexation agreements, it is not unusual to include specific provisions requiring the developer to construct and dedicate facilities beyond those characterized as local or on-site facilities. Such agreements may include a prorated portion of backbone facilities including WRRFs, lift stations, and larger-diameter interceptors. Once the facilities are dedicated, the utility would assume O&M expenses and replacement responsibility.

6.5 Leasing

6.5.1 *Capital Leasing*

Leasing, as a form of capital financing, is an arrangement that may allow a municipality to secure facilities in instances where other capital-funding mechanisms may not be readily available. Leasing is more widely used when acquiring equipment and facilities with a relatively short service life. Although leasing may be used as an alternative to debt financing, there are a variety of tax complexities to both the lessor and the lessee. A variety of tax laws can affect the benefit of leasing, and municipalities should thoroughly investigate and, as appropriate, obtain legal and tax assistance when considering any form of lease arrangement.

6.5.2 *Certificates of Participation*

Certificates of participation (COP) are types of leasing arrangements that can be used to finance capital improvements. They represent a way for wastewater utilities to finance capital projects without technically issuing long-term debt. Under a COP arrangement, an investor purchases a share of the lease revenue from a utility rather than investing in a utility's bonds. A leasing authority then leases an asset to the utility and transfers its right to receive lease payments to a trustee. The trustee then executes certificates of participation that are sold to investors. In this instance, the lessor is the governmental entity, and there may be any number of restrictions placed on the lessor regarding the use and disposition of the property financed with the COP. Certificates of participation can take many forms and have a variety of requirements, as dictated by state and local statutes. As with all financing alternatives, utilities should consult with financial and legal advisors when evaluating COP as a financing mechanism.

7.0 ISSUING MUNICIPAL SECURITIES: THE PROCESS

Having determined that some form of public financing (such as general obligation bonds or revenue bonds) is necessary to construct facilities contained in the CIP, the next step would be to assemble a financing team with the appropriate experience and expertise to approach the capital market. This process will involve not only municipal and utility managers, but also outside assistance from both legal and financial professionals.

The financing team may include the following:

- Parent government's financial staff,
- Utility staff,
- Independent auditor,
- Bond counsel,
- Disclosure counsel,
- Underwriter,
- Financial advisor,
- Financial feasibility consultant,
- Owner-advisor (PDB/P3/CMAR methods), and
- Engineering consultant.

A description of the functions and duties of each of these financing team members can be found in other publications (e.g., American Water Works Association, 2014). One change to the financing team occurred recently in response to the 2008 economic crisis and with the enactment of the resulting 2010 Dodd–Frank Wall Street Reform and Consumer Protection Act. This law created a new title of "registered municipal advisor" to address concerns that advisors of municipalities were providing important public financing advice without proper oversight. These advisory services include, among other things, guidance on the timing, sizing, interest rates, and structure of future debt. Typically, this role is covered by the financial advisor but could also include rate consultants and financial feasibility consultants under certain circumstances. To perform these services, the municipal advisor must be registered by the Securities and Exchange Commission and the Municipal Securities Rulemaking Board (MSRB) unless it is exempt under the rule. As a registered municipal advisor, the firm must establish and implement procedures to ensure compliance with MSRB rules and act in the best interest of the municipality that it is advising. Municipal advisor representatives must pass the Municipal Advisor Representative Qualification Examination (Series 50) to confirm their qualifications to provide municipal advice in compliance with the law and regulations. Furthermore, registered municipal advisors have a fiduciary duty to act in the best interest of the municipality that they are advising.

The first step in issuing municipal bonds is acquiring the authority to issue them. This involves demonstrating the financial viability of the project, defining the terms and conditions of the bond issue, attesting to its legality and tax-exempt status, and assessing the potential effect on its creditworthiness. The process of issuing municipal bonds varies depending on whether there is a public bidding process to select an underwriter. Typically, a public bidding process is followed to select an underwriter for the issuance of general obligation bonds, whereas a negotiated bidding process may be followed for the selection of an underwriter for the issuance of revenue bonds. Under a competitive bidding process, the financial advisor helps prepare the offering documents, bond counsel provides a legal opinion on the offering, the issuer prepares materials to present the financing plan to its governing body, and the governing body adopts a bond resolution authorizing the sale of the bonds. Under a negotiated bidding process, the underwriter is selected early in the process and helps prepare the offering documents, the issuer and the underwriter prepare materials to present to the governing body for approval, the governing body approves a trust indenture specifying the obligations of the issuer, and the governing body authorizes the sale of the bonds.

One of the most important tasks of the financing team is the development of an official statement (see the example in Figure 4.2). This document is a required disclosure document released by the municipal issuer that contains detailed information about the bonds and the financial condition of the issuer as well as the legal opinion verifying the municipality's authority to issue the bonds. It is the vehicle used by the issuing entity to present information to the underwriter and the investing public. It is initially prepared as a preliminary official statement and contains descriptions of the bonds to be offered and the way the proceeds are to be used. It also provides a summary of the utility and a discussion of operational and financial characteristics of the system, including the presentation of historical financial data. Other information such as the most recent audited financial statements, the financial feasibility and engineer's report, and the bond resolution and other legal documents may also be included as well. The preliminary official statement is initially distributed to prospective buyers and to underwriters in a competitive sale. After an underwriter is selected, the final official statement is printed.

Early in the debt-financing process, the issuer will make decisions regarding how the bonds will be sold. These methods include

- Private placement,
- Negotiated sale, and
- Competitive sale.

7.1 Private Placement

In a private placement, bonds are sold directly to a relatively small number of knowledgeable investors—investors who complete their own independent evaluation of the creditworthiness of the issuer. Private placements may be used by issuers likely to receive a noninvestment grade from a rating agency or issuers that desire to lower initial transaction costs or expedite the debt-financing process.

7.2 Negotiated Sale

In a negotiated sale, the placement of bonds is predetermined and an underwriting firm, or set of firms, is selected in advance of the sale. The issuer directly negotiates the price for the issue with the underwriting team at the time of the sale. Because of the complexities associated with evaluating the price offered by the underwriter, it is useful for the issuer to be assisted by other professionals, including a competent and experienced financial advisor who is registered as a municipal advisor. Issuers should consider their credit quality and tolerance for a public process when deciding between a private placement or negotiated sale. Also, in cases where the issue is a refunding of bonds or represents a new or unusual bond structure for the issuer, a negotiated sale may be preferred.

7.3 Competitive Sale

As the name implies, in a competitive sale, the issuer receives bids from underwriters or underwriting teams for the sale of bonds. After bids are evaluated, the issuer awards the sale to the underwriter or underwriting team that submits the lowest-interest costs. Some issuers are legally bound to hold a competitive sale as opposed to either a negotiated or private placement. With a well-run competitive process, the issuer is typically assured a competitive rate on the bond sale.

8.0 FINANCING CONSIDERATIONS FOR INVESTOR-OWNED UTILITIES

Much of the discussion up to this point has been applicable to both publicly, or municipally owned, and investor-owned wastewater systems. However, there are some additional considerations related to capital funding or financing that are pertinent to investor-owned systems because of the unique regulatory environment in which they operate (i.e., typically, investor-owned systems fall within the jurisdiction of state, public service, or public utility commissions). The regulatory authority of such entities encompasses a broad range of utility service issues, including the financing decisions of the utility.

In many jurisdictions, typically, an investor-owned wastewater utility must seek approval from the regulatory commission before the issuance of new debt. As a result, regulatory commissions have authority over the investor-owned utility's capital structure and debt levels. For example, typically, investor-owned utilities maintain a debt to total capitalization ratio in the range of 50% to 60%. A ratio higher than 60% may indicate that the utility is carrying too great a debt burden, whereas a ratio lower than 50% could be an indication of overuse of more expensive equity capital. This is one factor that a regulatory commission may consider in evaluating the proposed capital structure of an investor-owned wastewater utility as part of an application for new financing or a proposed rate increase.

Typically, the debt issued by most investor-owned utilities is taxable for the investor. Accordingly, taxable debt carries a higher interest rate for the issuer compared to tax-exempt debt. In addition, investor-owned utilities may be eligible for funds as part of state revolving loan fund programs because they are defined as eligible participants under the federal government's WIFIA program. In addition, investor-owned utilities may obtain financing from equity capital in the form of common and preferred stock. This form of financing is not available to municipally owned utilities because both common and preferred stock represent an ownership interest in the investor-owned utility. Common stockholders receive benefits in the form of share price appreciation and a portion of the utility's income in the form of discretionary dividend payments, whereas holders of preferred stock receive scheduled dividend payments, with a claim on income and assets that is senior to that of the common stockholders.

Investor-owned utilities recover capital-related costs in a different manner than many municipally owned utilities. For example, it is common for a municipally owned utility to recover capital-related costs, such as debt service and the portion of capital project costs funded on a pay-as-you-go basis through user rates established using the cash basis method of determining revenue requirements. However, investor-owned utilities often use the utility basis method, in which capital costs are recovered in the form of an annual provision for depreciation and as a return on rate base. Typically, system infrastructure contributed by developers is not included in the qualifying rate base and, as such, a rate of return is not earned on these assets by the investor-owned utility because they were obtained at no cost to the utility. Similarly, depreciation on contributed assets is also not included typically as part of the capital revenue requirement.

The original cost of the asset incurred by the investor-owned utility is returned over the course of the asset's depreciable life, and a rate of return is earned on the unrecouped cost. As a result, the depreciation and rate of return portions of the revenue requirement may be used by the utility to re-invest in the utility system. However, it should be noted that the total amount returned to the utility may be inadequate to fund the future replacement of the asset because of cost inflation. As a result, additional capital may be required for re-investment in the utility system, especially during periods where substantial cost inflation has occurred. Furthermore, as rate base assets depreciate, their undepreciated values decline, which, in turn, results in a lower dollar amount earned as a return from each asset. Therefore, continued investment to the system's rate base helps to maintain the level of funds earned from the return on rate base and may reduce the need for external funding sources if these revenues are used for asset replacements.

9.0 SUMMARY

The financing of wastewater system capital improvements involves identifying current and future system needs and selecting funding and financing options that best align with the utility's goals. Adhering to established capital-funding policies can streamline the process. Wastewater utilities have access to a range of internal and external funding options, including state and federal grants and low-interest loans. The chosen funding mix should align with utility goals and policies, considering its effect on customer rates and overall financial health. When considering long-term debt issuance, utilities should assess their legal authority and engage a qualified team for bond sales.

10.0 REFERENCES

American Water Works Association. (2014). *Water utility capital financing, manual of water supply practices M29* (4th ed.).

Water Research Foundation. (2016). *New and emerging capital providers for infrastructure funding* (Project No. 4617).

5

Determination of Revenue Requirements

1.0 INTRODUCTION

To provide adequate service to its customers, every wastewater utility must receive sufficient annual revenue to ensure proper operation and maintenance (O&M) of facilities, development and perpetuation of the physical condition of system assets, compliance with regulatory requirements, and maintenance of the utility's financial integrity. The first critical step in setting rates necessary to generate adequate revenue levels is determining system revenue requirements.

A financially sustainable utility must address several issues when developing its revenue requirements. First, the utility must identify the appropriate approach to be used in establishing its revenue requirements. Second, it must select the appropriate representative year within the planning period, sometimes termed the "test year," to be used in the allocation of costs of service to the classes of customers served by the utility and for the design of rates or user charges to recover allocated costs.

2.0 APPROACHES TO DETERMINE REVENUE REQUIREMENTS

The two generally accepted and practiced approaches for projecting total revenue requirements of a wastewater utility are the "cash-needs" (or simply the "cash") approach and the "utility-basis" approach. Each has a proper place in utility rate-making practice and each, when properly used, can provide for sound utility financing and cost recovery from the utility's customers. Factors affecting the selection of the approach to determine revenue requirements include the ownership of the utility (public or municipal vs. investor), state regulatory jurisdiction regarding utility rates, and whether the utility provides service to retail or wholesale customers located outside of the political boundaries of the utility. Table 5.1 contrasts the key differences between the cash-needs approach and the utility-basis approach. Municipal utilities in the United States most prevalently use the cash-needs approach because the approach parallels municipalities' budgeting process. Given its popularity among municipal utilities, the cash-needs approach is described in detail in the following sections.

The utility-basis approach is commonly used by privately owned and regulated utilities. Under the utility-basis approach, revenue requirements are calculated using O&M expenses, annual depreciation, property taxes in various jurisdictions, or payments in lieu of taxes (PILOT), and an appropriate level of annual return on rate base. There are several reasons why municipal utilities can or should use a utility-basis approach for their cost-of-service allocations, including wholesale rate-making, determining

TABLE 5.1 Comparison of Approaches to Determine Revenue Requirements

Cash-needs approach	Utility-basis approach
O&M expenses	O&M expenses
Debt service payments	Depreciation expense on rate base assets
Rate revenue funded capital expenditures	Return on rate base
Change in cash reserves	
Total revenue requirement	**Total revenue requirement**

outside-city rate differentials, or better alignment with budgeting process. The utility-basis approach is covered in detail in Appendix A.

3.0 CASH-NEEDS APPROACH

The essence of the cash-needs approach is that revenues of the utility must be sufficient to recover all cash requirements over the utility's financial-planning period. The cash-needs approach is basically an extension of cash-oriented budgeting and accounting systems are used by municipally owned utilities and, accordingly, typically, is more readily understood by the governing bodies of these organizations. In the context of determining a utility's revenue requirements, "cash needs" should not be confused with "cash" in accounting terms. The accounting term "cash" refers to recognizing revenues when cash is received and expenses when payments are made, whereas "cash needs" pertains to the total revenue required to cover all annual cash expenditures and any additional cash needed to meet any financial policies and/or obligations of the utility. The cash-needs approach in determining revenue requirements may be evaluated by the utility on either a cash basis or an accrual basis of accounting. For example, operating expenses may be projected using prior year actuals or budget, and capital expenditures may be projected using appropriations, encumbrances, or cash needs. Practitioners should be careful to consider the potential tradeoffs with each of these approaches and the effects they may have on fund balance and other financial metrics.

The cash-needs method excludes depreciation and includes principal repayment (and interest) on outstanding debt. The main objective in using the cash-needs method is to collect revenues to cover cash expenses. Revenues must cover annual expenses and reserve needs for capital and operational funds as well as meet financial obligations such as debt service coverage. Inclusion of depreciation is not appropriate in developing cash-needs revenue requirements. Annual depreciation expense is a noncash expense that represents the amortization of historical capital expenditures. Because the cash-needs approach uses principal and interest on debt plus routine capital expenditures (as expensed or encumbered), including depreciation would result in double counting of capital costs. It should be noted that depreciation can have a place in cash-needs rate-making; for example, it is commonly used for setting capital reserves targets or establishing a benchmark for the minimum level of normal annual system replacements and improvements.

Municipally owned utilities must balance their revenues and expenses within a cash budget framework. Unlike investor-owned utilities, municipally owned utilities do not have access to sources of capital other than retained earnings, grants, and formally issued debt typically. Therefore, the total revenues collected from all customers must equal expenses plus any additions to reserves. Reserves are often established for O&M, working capital, and debt service needs. Positive year-end balances may be used to accumulate reserve funds. The desirable level of reserve funds is an important item in the development of a financial plan (see Chapter 3).

The balancing of revenues and cash expenses for the current period is the foundation for the cash-needs approach to developing revenue requirements. Common cash-needs revenue requirement components include the following:

- O&M expenses,
- Normal annual system replacements and improvements,
- Equipment purchases,
- Cash-financed significant capital improvements,
- Debt service payments,
- Additions to reserves,
- Debt service coverage,
- Payments in lieu of taxes, and
- General fund transfers.

In projecting utility revenue requirements, an analysis of historical data, information, and trends often provides a useful tool or guidance in this process. Historical data must be normalized or adjusted to reflect conditions that may not continue into the future. Such factors include, but are not limited to, the following:

- Number of customers served,
- Customer water usage adjusted for water not discharged to the wastewater system,
- Weather,
- Inflation,
- Interest rates,
- Capital-financing needs, and
- Economic conditions.

3.1 Operation and Maintenance Expenses

Typically, O&M expenses are projected based on actual historical expenses or budgeted expenses and are adjusted for anticipated changes in expenditures during the projection period. Typically, these expenses are categorized according to a uniform system of accounts relevant to the utility, which may be state mandated. Typically, O&M expenses include salaries and wages, fringe benefits, purchased power, other purchased services, chemicals, other materials and supplies, expenditures for replacement of small equipment items, and direct and indirect overheads. For a municipally owned utility, overhead elements of O&M expense may include the costs of support service provided to the utility (typically by a general fund or other municipal fund), such as legal, personnel, fleet management, information technology and computer systems support, and billing services. Such expenditures are properly included in utility revenue requirements, just as they would be included if such services were to be provided directly by the utility itself.

Table 5.2 shows an example projection of O&M expenses for a 5-year study period. Often, a utility's current budget provides a reasonable starting point for such projections; in Table 5.2, the first column of figures is noted as the current year budget. In the example, an allowance for inflation of 3% is reflected in the projections across all O&M expense categories. Detailed analysis of the potential future inflation for a particular utility may indicate that various components of O&M expenses are expected to experience differing rates of inflation. For example, electric power, chemical, and employee healthcare and retirement costs may increase at higher rates than other categories of O&M expenses. The development of O&M expense projections also requires the utility to consider how changes in factors such as customer discharge volumes, customer account growth, treatment processes, biosolids processing, or regulatory requirements may cause certain O&M expenses to deviate from historical norms. These factors are often a key reason why a utility may choose to develop O&M expense projections that feature a percentage change in expenses that differs from actual historical experience. Table 5.2 presents an allowance for additional costs associated with the expansion of the water resource recovery facility (WRRF) as an example of how the considerations above can be incorporated to O&M expense projections. As indicated in the capital program financing discussion included in Table 4.1, a significant WRRF expansion project is expected to be completed in year 3 of the projection period. Accordingly, in year 4 of the projections shown in Table 5.2, 12% increases in staffing, power, chemicals, and other expenses are recognized in both the wastewater treatment and sludge disposal expense because of this facility expansion.

Table 5.3 has been prepared to accumulate the overall example projected revenue requirements on a cash basis. The first component of revenue requirements shown in Table 5.3 is O&M expenses. A summary of projected O&M expenses is show on lines 1 through 5 of Table 5.3.

TABLE 5.2 Example Projection of O&M Expenses for a 5-Year Study Period

	Current Year Budget	Year 1	Year 2	Year 3	Year 4	Year 5
Wastewater treatment						
Personnel	$1,754,000	$1,807,000	$1,861,000	$1,917,000	$2,175,000	$2,240,000
Power	$702,000	$723,000	$745,000	$767,000	$870,000	$896,000
Chemicals	$350,500	$361,000	$372,000	$383,000	$434,000	$447,000
Other expenses	$702,000	$723,000	$745,000	$767,000	$870,000	$896,000
Sludge disposal						
Personnel	$1,842,000	$1,897,000	$1,954,000	$2,013,000	$2,323,000	$2,393,000
Power	$737,000	$759,000	$782,000	$805,000	$929,000	$957,000
Chemicals	$369,000	$380,000	$391,000	$403,000	$465,000	$479,000
Other expenses	$737,000	$759,000	$782,000	$805,000	$929,000	$957,000
Industrial waste	$1,030,000	$1,061,000	$1,093,000	$1,126,000	$1,160,000	$1,195,000
Subtotal treatment	$8,223,500	$8,470,000	$8,725,000	$8,986,000	$10,155,000	$10,460,000

Wastewater transmission						
Pumping and lift stations	$474,000	$488,000	$503,000	$518,000	$534,000	$550,000
Equipment maintenance	$726,000	$748,000	$770,000	$793,000	$817,000	$842,000
Supervision	$430,000	$443,000	$456,000	$470,000	$484,000	$499,000
Subtotal transmission	$1,630,000	$1,679,000	$1,729,000	$1,781,000	$1,835,000	$1,891,000
Wastewater collection administration	$3,456,500	$3,560,000	$3,667,000	$3,777,000	$3,890,000	$4,007,000
Billing and collection	$1,076,500	$1,109,000	$1,142,000	$1,176,000	$1,211,000	$1,247,000
Office of the director	$473,000	$487,000	$502,000	$517,000	$533,000	$549,000
Engineering	$749,500	$772,000	$795,000	$819,000	$844,000	$869,000
Planning and project management	$944,500	$973,000	$1,002,000	$1,032,000	$1,063,000	$1,095,000
All other expenses	$603,000	$621,000	$640,000	$659,000	$679,000	$699,000
Subtotal administration	$3,846,500	$3,962,000	$4,081,000	$4,203,000	$4,330,000	$4,459,000
Total O&M	$17,156,500	$17,671,000	$18,202,000	$18,747,000	$20,210,000	$20,817,000

TABLE 5.3 Overall Example of Projected Revenue Requirements for a Cash-Needs Approach

	Current year budget	Year 1	Year 2	Year 3	Year 4	Year 5
Operation and maintenance expense						
Wastewater treatment	$8,223,500	$8,470,000	$8,725,000	8,986,000	10,155,000	$10,460,000
Wastewater transmission	$1,630,000	$1,679,000	$1,729,000	1,781,000	1,835,000	$1,891,000
Wastewater collection	$3,456,500	$3,560,000	$3,667,000	3,777,000	3,890,000	$4,007,000
Administration	$3,846,500	$3,962,000	$4,081,000	4,203,000	4,330,000	$4,459,000
Total O&M expense	$17,156,500	$17,671,000	$18,202,000	$18,747,000	$20,210,000	$20,817,000
Operation and maintenance reserve	$62,000	$64,000	$66,000	$68,000	$183,000	$76,000
Capital requirements						
Debt service—existing bonds	$5,230,000	$5,230,000	$5,230,000	$5,230,000	$5,230,000	$5,230,000
Debt service—proposed bonds	—	$2,800,000	$2,800,000	$2,800,000	$2,800,000	$2,800,000
Equipment purchases	$547,000	$563,000	$580,000	$597,000	$615,000	$633,000
Capital improvements from revenue	$2,500,000	$3,000,000	$1,000,000	$3,000,000	$2,000,000	$3,000,000
Total capital requirements	$8,277,000	$11,593,000	$9,610,000	$11,627,000	$10,645,000	$11,663,000
Payments in lieu of taxes	$920,000	$950,000	$975,000	$1,005,000	$1,035,000	$1,065,000
Total revenue requirements	$26,415,500	$30,278,000	$28,853,000	$31,447,000	$32,073,000	$33,621,000
Less other nonrate revenue	($15,000)	($16,000)	($17,000)	($18,000)	($19,000)	($20,000)
Total revenue requirements from rates	$26,400,500	$30,262,000	$28,836,000	$31,429,000	$32,054,000	$33,601,000
Annual debt service coverage (times)	1.77	1.57	1.33	1.58	1.48	1.59

3.2 Equipment Replacement

Equipment replacement for many utilities occurs on a somewhat irregular basis, with little to no spending for many years and then one large expenditure in other years. Some utilities simply accept the spike in the spending as part of their long-term financial-planning process. However, many utilities, especially for smaller systems, equipment replacements can be a material effect to the fund balance. Establishing consistent revenue requirements for equipment replacement costs may be an essential undertaking. A potential methodology for determining an appropriate annual level for equipment replacement cost is as follows:

1. Identify all equipment within the wastewater utility with an average life expectancy shorter than that of the overall life expectancy of the WRRF structures,
2. Estimate the cost of replacing the identified equipment in the previous step and the estimated replacement time intervals, and
3. Calculate the amount to be reserved annually to pay for the replacement of the equipment when necessary.

It should be noted that a detailed financial plan for establishing and building reserve funds for equipment replacement costs of both collection systems and WRRF equipment should be formalized. With such a plan, significant future user charge effects can be mitigated. The implementation of long-term asset management systems can greatly improve the projection of needed capital expenditures to maintain fixed assets adequately. An asset management plan, combined with sufficient and long-term funding through user charges, should significantly improve utility management's ability to maintain utility assets over time. Additionally, federal regulations require that federal grant-funded wastewater systems must recover at least O&M expenses and equipment replacement costs through their system of user charges. Examples include regulations enforced by the U.S. Environmental Protection Agency and the U.S. Department of Agriculture's Rural Utilities Service. Other than equipment replacement requirements, these regulations do not prescribe the manner in which capital costs must be recovered.

3.3 Capital Expenditures

The establishment of cash requirements for future facility improvements or additions is a decision typically affected by the nature of investment and utility financing policies. Capital expenditures are generally classified into one of two categories: normal annual (routine) replacements and significant capital replacements and improvements. A utility should periodically review and update its needs in each of these areas to recognize changing conditions. Projections of such needs are an essential part of overall revenue requirement development. Total capital needs projections should be accompanied by estimates of contributions from sources such as developer contributions, system development fees, grants, and other nonutility sources. Chapter 4 of this manual discusses these funding sources.

It is common practice for municipally owned utilities to fund normal annual replacements and improvements from current revenues. Typically, such expenditures cover ongoing expenditures for smaller mains and replacement of other existing assets and are consistent in amounts each year, though not always.

Typically, significant capital expenditures are financed through the issuance of long-term debt, which better matches the term of the financing with the life of the facility being financed and helps to spread the costs of the financing over future users who benefit from the facilities. Note that utilities may choose not to exclusively use debt to fund significant capital expenditures. When funding capital expenditures or portions thereof with cash, prudent management of financial reserves is critical to address volatility in cash capital needs year-over-year. This must be accounted for in the revenue requirements through contributions to/from reserves, as discussed in Section 3.6.

Lines 9 and 10 of Table 5.3 show projected revenue requirements for other capital expenditures. Capital improvements from revenues, shown on line 10, are derived from Table 4.1 of this manual.

3.4 Debt Service Payments

The debt service component of the cash-needs revenue requirements approach consists of annual principal and interest payments during the study period on existing bonds or other debt instruments. The amount of cash required to pay principal and interest on outstanding debt may be obtained from established debt service payment schedules. In addition to the debt service on existing bonds, the debt service on any additional bonds that are projected to be issued during the study period should also be recognized in revenue requirement projections. Examples of debt service on existing bonds is shown in Table 5.3.

As discussed in Chapter 4 and presented in Table 4.1, it is expected that \$43 million of the costs of the WRRF expansion project will be financed through the issuance of bonds in year 1 of the projection period. Assuming an annual interest rate of 5.0% on the proposed bonds, an equal annual principal, and interest payment schedule over 30 years, the annual debt service payment on the proposed bond issue amounts to \$2,800,000 per year as shown in Table 5.3.

In addition to the payment of principal and interest on existing or proposed bonds as discussed in Chapter 4, typically, revenue bonds also have bond covenants that dictate the specific structure of the debt-financing mechanism used for funding and require the establishment of a debt service reserve fund. Typically, the level of debt service reserve fund is established at 1 year's principal and interest payment or the maximum future year's principal and interest payment on the cumulative total of all outstanding revenue bonds. Typically, this debt service reserve can be funded in three ways:

1. From the bond proceeds, which would require an increase in the principal amount of a particular bond issue. In this case, the total amount borrowed would need to be sufficient to cover construction costs plus the costs of issuing the bonds and the debt service reserve amount.
2. From annual system revenues, typically over a 5-year period.
3. Through the purchase a surety bond insurance policy. The insurance policy payment becomes a part of the bond issuance expense. Note that surety bond insurance is used for the example in Table 5.3.

3.5 Other Cash Requirements

Other cash revenue requirements that may need to be financed from utility system revenues include payments to the general fund of the municipal government owner. Courts in several states have confirmed the right of municipal utilities to earn a rate of return from the provision of utility service that can be used to provide funds for payments to the general fund. Depending on the unique local circumstances of the utility in question, these payments can take the form of PILOT, gross receipts taxes, or a dividend payment calculated using a variety of methodologies. It is recommended that utility governing bodies establish formal policies regarding the scope, calculation methodology, and limits of such transfers. A PILOT is shown in the example revenue requirement projection presented in Table 5.3. Typically, this payment is based on a measure of the property value of the utility, which would be payable to the municipality if the utility was investor-owned.

3.6 Reserve Fund Management

As discussed in Chapter 3, there are many types of reserves and reserve funds that may be maintained by the utility. The most common types of reserves are as follows:

1. General operating reserves,
2. Emergency capital reserves,

3. Rate stabilization reserves,
4. Bond covenant reserves, and
5. Insurance reserves.

Most utilities need a general operating reserve (O&M reserve) to manage unforeseen fluctuations in expenses. This reserve is often in the range of 30 to 90 days (1 to 3 months) of operating costs. For purposes of the O&M revenue requirements summarized in Table 5.3, it is assumed that the example utility must maintain an O&M reserve equal to 45 days' (or approximately 1.5 months') worth of annual O&M expense. It is assumed that, in the current year, the example utility has a fully funded O&M reserve. Accordingly, by the start of each year in the projection period, the utility must add to the O&M reserve an amount equal to the incremental increase in annual O&M expense from the previous year. The utility may either build up the reserve during the previous year at a self-determined incremental rate or by a year's end addition of a full increase from uncommitted fund balances. An O&M reserve equivalent to 45 days of total O&M expenses may not be adequate for some utilities. Utilities should evaluate their financial needs and adjust the level of O&M reserve that they need to maintain and include it as a financial performance metric in the utility's financial policies.

Another common reserve for utilities that issue revenue bonds is the bond covenant reserve. Often, the bond resolution associated with the issuance of revenue bonds or other types of bonds stipulates that the utility maintain certain reserve fund balances, typically which include a year's worth of debt service payments.

Reserves can be designated as restricted or unrestricted. A restricted reserve is for a specific purpose and associated funds can only be used for that purpose. An unrestricted reserve offers more flexibility because it can be used for any purpose that the utility desires.

Chapter 3 presents a comprehensive discussion of reserves, including their purpose, funding, and use. Financial reserves play an important role in prudent utility financial management.

3.7 Debt Service Coverage

Debt service coverage requirements associated with most debt issuances provide for the generation of revenues over and above annual principal and interest payments on the bonds, which may be used for cash financing of a portion of the significant capital improvement program.

Most revenue bond ordinances include specific requirements related to annual debt service coverage. Such requirements are for the protection of the purchasers of the bonds in that they require the utility to establish rates that generate revenues in excess of annual debt service payments. In general, the formula for the calculation of debt service coverage is as follows: *net operating income* defined as total system revenues (potentially with some exclusion for certain types of interest income) less *O&M expenses*. The specific items included in O&M expenses for the purposes of calculating debt service coverage may go beyond those associated typically with the actual cost of operating and maintaining the utility system. For example, the utility bond indenture may specify that PILOT, gross receipts taxes, or other forms of payments to the general fund of the municipal owner be included within O&M expenses to calculate net operating income as used in the debt service coverage calculation. Typically, O&M expenses, as used in debt service coverage calculations, do not include depreciation or amortization expenses. *Debt service coverage* is defined as annual net operating income divided by the total of annual debt service payments, expressed as a decimal number or a percentage.

The resulting legally required minimum debt service coverage is generally established between 1.10 and 1.25 times coverage in most bond ordinances. This "extra" revenue, amounting to between 10% and 25% of annual debt service payments, provides a "cushion" for the utility in meeting its O&M expense and annual debt service requirements. Generally, annual debt service coverage applies only to the senior, or parity, debt of the utility, as shown in the last line of Table 5.3. Typically, subordinate bonds or other debt instruments require a lower coverage of revenues after providing for O&M expenses and the payment of senior bond debt service.

The coverage cushion provided by legally required minimum debt service coverage may be necessary in the event of a shortfall in annual revenues because of climatological, economic, or other factors. The "extra" revenue, however, is also available to the utility for funding various revenue requirements, including reserves, payment of subordinate debt service, the financing of normal annual or significant capital expenditures, and other legal uses, which are defined in the revenue bond ordinance.

Many utilities follow a financial-management policy of planning for and implementing rate increases designed to achieve a target level of debt service coverage in excess of the legally required minimum. For example, a utility with a minimum debt service coverage requirement of 1.25 may choose to implement rates that will achieve 1.50, 1.75, or even 2.0 times debt service coverage. Utilities that use a policy of setting debt service coverage targets in excess of the legally required minimum must carefully weigh the customer bill effects of charging higher rates against the financial-management benefits that this strategy can provide. One obvious benefit is that the utility will maintain an even larger coverage cushion to meet its contractually obligated debt service payments. Bond rating agencies often consider a large coverage cushion to be a positive factor when assessing the creditworthiness of the utility. Moreover, in some circumstances, a higher debt service coverage ratio may lower the utility's borrowing cost if the utility receives a higher bond rating.

3.8 Revenue Requirement Offsets

Utilities earn revenues from a variety of sources other than the rates paid by customers for wastewater service. These miscellaneous sources of nonrate revenues can include lease and rental revenues, interest income earned on cash or investment reserve balances, miscellaneous customer service fees and penalties, and receipt of system development charges (see Chapter 10) paid for new connections to the utility system. Increasingly, utilities are also earning nonrate revenues from sources such as the sale or resale of wastewater treatment byproducts, including biosolids, recycled water, and electric power. All of these sources of nonrate revenues can be used to reduce the test year revenue requirement that must be paid by customer rates.

3.9 Total Revenue Requirements From Rates

Total revenue requirements can be calculated in three key steps. First, calculate cash needs by taking the sum of all operating expenses, equipment purchases, cash funded capital, and debt service on existing and proposed bonds and loans. Then deduct any nonrate revenues (as defined in section 3.8). Next, project annual rate revenues. Finally, evaluate the need for any additional revenues to meet debt service coverage, reserve targets, or any other financial goals and objectives. The combination of the projected total operating and capital expenditures, use of nonrate revenues, and any additional cash needed to meet financial policies and obligations comprise the total revenue requirement for your enterprise.

4.0 THE COST-OF-SERVICE PROCESS AND DETERMINING A TEST YEAR

The first step in the cost-of-service process is deciding on the approach to be used for determining revenue requirements (either the cash-needs or utility-basis approach). Once an approach for revenue requirements is established, the next step is determining what year or period should be selected for purposes of allocating the cost of providing service to customer classes and designing rates to recover these allocated costs. The selection of the specific year or period of years for such analyses is triggered typically by the need for an overall adjustment in revenue levels. Although it is not inappropriate to look at the cost justification and equitability of a utility's rate structure even when no revenue increases are indicated, such analyses most often accompany an overall revenue-level increase. In some situations, there are regulatory procedures that govern the selection of the period to be analyzed for rate adjustments. This section provides guidance for the selection of the period for cost-of-service allocations and rate design.

4.1 Test Year

When determining revenue requirements for cost-of-service allocations and rate design, typically, a utility should select an appropriate "test year." This is customarily a 12-month period of time that is chosen because it has annual expenditures that are representative of the year(s) during the period for which the new rates are expected to be in effect. A single or multiyear period can be selected for rate setting. The multiyear alternative might be useful when the utility is expected to experience significant changes in its cost structure over a specified period of years or when phasing in a rate structure to help even out large increases or decreases in rates from year-to-year.

Typically, most municipally owned utilities choose a forward-looking or future test year. Use of a future test year enables the utility to set rates prospectively to meet budget planning and bond indenture requirements. This provides more flexibility to the utility and avoids the situation of always "playing catch up," which is the case typically when using a historical test year for rate-setting.

Typically, investor-owned utilities are under strict guidelines and must follow practices and regulations established by the public service commissions or other regulatory agencies. These practices, which vary by commission, dictate whether the investor-owned utilities must use a historic, current, or future test year in determining revenue requirements. Often, the pro forma test year revenue requirement for investor-owned utilities is based on the most recent year's audited cost data modified to recognize known and measurable changes for the future year. Such known and measurable changes might include newly adopted tariffs by the utility's electric power provider, newly negotiated labor union contracts, changes in chemical prices, and similar changes in the utility's revenue requirements that were not reflected in the most recent year of record.

In some states, retail and/or wholesale rates charged by municipal utilities are regulated by state public service commissions. Similarly to investor-owned utilities, municipal utilities must submit evidence to the state public service commission justifying the need for additional revenues. Despite this similarity, the strict guidelines that investor-owned utilities must follow when determining their test year revenue requirement may not, in all cases, be applied to municipal utilities. For example, a public service commission may accept a municipal utility's proposed test year revenue requirement as developed during the municipal budgeting process without the need for strict adherence to the policy that the test year revenue requirement be based on known and measurable changes as applied to the most recent year's audited cost data. In such cases, it may be sufficient for the municipal utility to provide evidence describing why budgeted expenses have increased and how actual expenses have closely matched budgeted expenses in past years.

4.2 Establishing Revenue Requirements for Cost of Service

Once an approach and test year/period for determining revenue requirements are established, the next step is calculating the actual revenue requirement for cost of service and rate-making. For purposes of presenting a cost-of-service allocation and rate design example (which are discussed in Chapters 6 through 8 of this manual), it is determined based on a review of the utility's revenues under existing rates and its revenue requirement projections shown in Table 5.3 that an overall adjustment in the level of existing revenues and rates is required for year 2 of the planning period. The Year 2 column in Table 5.3 summarizes total test year revenue requirements using the cash-needs approach.

5.0 CONCLUSION

Annual revenue requirements methods vary depending on ownership, regulatory requirements, local policies, and local circumstances. Because most municipally owned wastewater systems are budgeted on a cash basis, the most common method for determining revenue requirements is the cash-needs approach. However, either the cash-needs or the utility-basis is appropriate for use by municipally owned utilities in determining annual revenue requirements. When the utility is regulated or serves outside users, the utility-basis is commonly used to ensure equitable and defensible distribution of cost responsibilities to all system users.

The total revenue requirement from rates defines the costs that must be recovered from customers for the provision of utility service. These costs are then allocated to customer classes based on the unique demands they impose on the utility system (see Chapters 6 and 7). Rates are then designed to recover the allocated customer class cost of service (see Chapter 8).

6

Allocating Costs of Services to Cost-Causative Components

1.0 INTRODUCTION TO COST ALLOCATIONS

The cost-allocation process determines the cost (revenue requirement) of providing wastewater service to each of a utility's customer classifications. This process includes the functionalization and allocation of wastewater system revenue requirements (the cost of service), discussed in this chapter, followed by the distribution of costs by customer classification based on the flow, capacity, customer-related costs, and strength for which each class of service is responsible (see Chapter 7). The customer class cost-of-service information resulting from this process will serve as a useful tool to the utility in providing equity in its rate structure.

For wastewater utilities, costs generally fall into three broad types: (1) fixed (e.g., capacity, customer, or billing); (2) flow; and (3) extra strength (costs of managing and treating wastes substantially in excess of those for domestic flows). A utility's individual circumstances may warrant additional cost component categories. Depending on a utility's objectives and risk profile, revenue stability may be a concern and can be addressed by assigning more costs to fixed cost components.

The cost-of-service allocation process determines the portion of net revenue requirements created by serving each customer class. Typically, rates that reflect the costs of providing wastewater service to the different customer classes are viewed as fair and equitable; deviations from cost of service may be warranted to advance specific community policy objectives, but, in general, should be limited and fully disclosed. Figure 6.1 shows how these elements come together to determine recommended rates.

1.1 Steps to Cost Allocation

Chapter 6 provides definitions and examples of key cost-allocation terms and discusses two distinct methodologies (design basis and functional basis) used in the wastewater utility industry for allocating annual revenue requirements to cost components.

Regardless of the methodology being applied, the cost-of-service process includes the following steps, which are graphically displayed in Figure 6.2:

1. Assign annual revenue requirements to function or activity. Typically, these functions include collections, liquid and solid treatment processes, sludge processing (which often produces biosolids), and so on, but can be anywhere from a high level to the process level to whatever level of detail the utility desires (see Chapter 5 for discussion on revenue requirements. See Sections 3.0 through 5.0 of this chapter for discussions about allocations of costs to functions or activities);

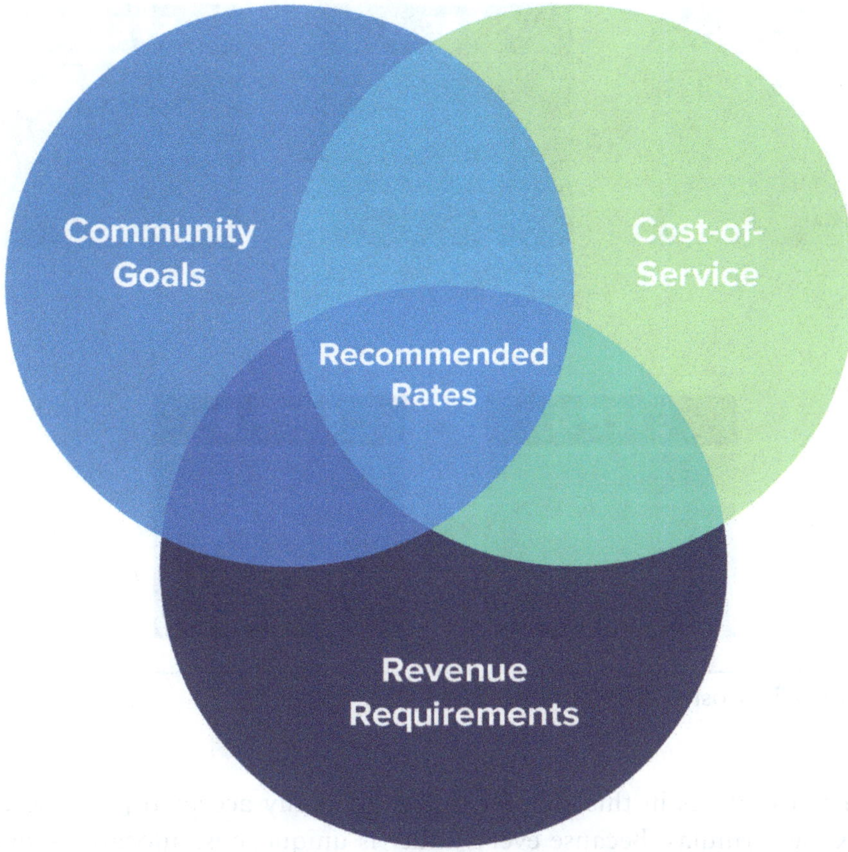

FIGURE 6.1 Melding of Components to Arrive at Recommended Rates

2. Allocate these functional costs to appropriate cost-causative components, including those related to flow, capacity, inflow and infiltration (I/I), and effluent strength (e.g., biochemical oxygen demand [BOD], total suspended solids [TSS], or other parameters), and others as appropriate to the system (see Sections 6.0 through 10.0 of this chapter for discussions about allocating functional costs to cost-causative components);
3. Develop units of service by customer class for each cost component (see Chapter 7);
4. Develop unit costs of service by dividing the total costs for each cost component by the respective total system units of service (see Chapter 7); and
5. Distribute costs to customer classes based on the unit costs of service and each class's units of service for each cost component (see Chapter 7).

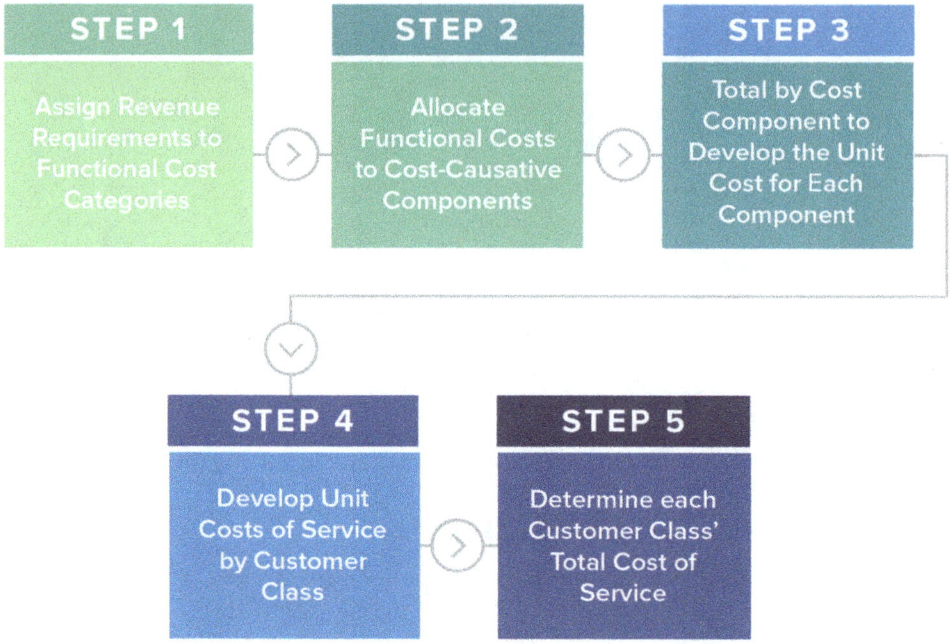

FIGURE 6.2 Cost-of-Service Flowchart

The examples in this chapter reflect generally accepted processes, not inflexible formulas. Because every utility is unique, cost allocations should be tailored to and represent appropriate cost-causative factors for each local situation as well as the financial goals and objectives of the specific utility. Utilities are heterogeneous in critical ways, including their size, technical complexity, treatment processes, service areas, customer demand patterns, collection system dynamics, or importance of industrial loadings, among other factors. Differences such as these create material distinctions that must be recognized for cost-allocation processes to reflect locally relevant conditions. The size of the utility or water resource recovery facility (WRRF) might seem simple to comprehend and recognize but cost differences do not necessarily vary in linear or even particularly predictable fashion. Cost-allocation processes may differ in the face of high- and low-strength waste streams, seasonal and steady annual demand patterns, new versus older facilities (especially if older facilities may have been repurposed to address the evolution of treatment processes resulting from regulatory change), or the presence or absence of a wide variety of resource recovery technologies and processes. As a result, utility managers and rate study teams must develop an understanding of the unique aspects of the utility in question before seeking to apply practices broadly applicable in the industry.

When allocating costs to cost components, wastewater utility management should consider the nature of the service provided and the use of facilities by the various customer classes the utility serves. This evaluation helps determine the types of cost-causative factors present in a utility's operations and appropriate considerations for establishing rates. To this end, it is useful to consider the following cost-causative factors: directly allocable, common-to-all (meaning the costs that should be shared among all users of the system, whether retail or wholesale, inside the city or outside the city), seasonal, and growth-related cost assignments, along with volume, peak rates of use, influent wastewater strength, I/I, and customer-related costs, which are discussed more in this chapter.

1.2 Timing

To ensure that the utility's rates remain fair and reasonable, utility management should periodically conduct a cost-of-service rate study that includes all of the elements of cost allocations. Typically, cost-of-service analyses should be conducted every 3 to 5 years if utility operations, cost factors, and service conditions are stable. If cost-of-service studies are not conducted on a regular basis, at a minimum, one should be conducted in advance of any material change, such as when large capital projects are planned and/or significant new users join the system. Either of these types of changes can affect cost allocations and require an updated cost-of-service analysis to ensure the maintenance of equitable rates and charges.

2.0 TYPES OF WASTEWATER UTILITY SERVICE FUNCTIONS AND COSTS

Typically, a wastewater utility consists of component facilities to collect wastewater from utility customers by use of small collection sewers, transport it by gravity or lift station pumping to larger intermediate pumping stations, and further pump it through force mains or gravity-feed it through large interceptor sewers to WRRFs. At such facilities, wastewater is treated by settling and biological and/or chemical processes to reduce its organic and chemical strengths to specified levels, permitting it either to be safely discharged through outfall sewers to receiving waterways via deep wells and/or infiltration basins for groundwater recharge or reclaimed for non-potable use or indirect or direct potable use. Treatment processes result in residual matter that must be further processed before it may be converted to a fertilizer or other beneficial product, incinerated with ash disposal to an approved landfill, or transported to landfills to be deposited in or used as solid waste cover material, as permitted. Different treatment processes, customer profiles, and environmental conditions lead facility designers, operators, and managers to focus on different functions.

3.0 COST-ALLOCATION PROCESS

Cost allocation is the process of apportioning costs into various cost or activity centers, and then allocating those costs of service to the appropriate cost components for subsequent distribution to customer classes based on their service requirements. The process provides a sound basis for establishing rate schedules that equitably recover costs from all customer classes by recognizing differences in the costs of providing service to customers with different characteristics.

Although some costs may be directly assigned to certain classes of service, most utility costs must first be apportioned to cost components based on cost causation or functional considerations. A utility with multiple subsystems may need to add in extra functionalization. Costs may have to be allocated based on percentage use of investment; operation and maintenance (O&M) labor; or chemicals, materials, supplies, and power cost components for each subsystem (e.g., differences in solids-handling processes across facilities). If multiple facilities or subsystems have identical purposes, there is no need to apportion costs differently.

The utility incurs annual operating expenses and capital costs for each of its component facilities and service functions and must fully recover these costs, regardless of whether the facilities or functions are used for one day or every day of the year. The benefit (the opportunity to use the facilities) exists whether the facilities are actually used or are merely there as a backup.

4.0 COST OR ACTIVITY CENTERS

At the most basic level, wastewater utility cost or activity centers include wastewater collection and conveyance from customers, pumping, interceptor transportation, treatment, effluent disposal, residuals disposal, and customer-related services. Cost centers represent individual or grouped activities for which the utility's chart of accounts should permit the capture of direct and indirect O&M expenses and capital costs related to operating the utility, as discussed in Chapter 5. By extension, cost centers reflect the smallest significant units of management responsibility for all system assets and operational activity associated with each wastewater utility function and service.

Typically, the number of cost or activity centers varies by utility, depending on whether the basis of allocation is operations or facility investment, as shown in Table 6.1.

For utilities that have established their accounting records according to recommendations of the National Association of Regulatory Utility Commissioners' charts of accounts, revenue requirements will already be expressed by the aforementioned utility functions and cost or activity centers. For other utilities that operate under a different account structure or do not have this level of detail established, revenue requirements should be allocated to such cost centers at the beginning of the cost-allocation process to the extent practical. The utility should associate existing budget categories with the most relevant allocable cost centers. This approach may provide the utility with a reasonable allocation of budget costs to defined cost centers; however, shifting toward a structure where budget categories are more easily and closely aligned to functional cost centers may better support the development of cost-based rates.

5.0 COST COMPONENTS

Cost components represent the key operational and service elements of a wastewater utility to which separate cost groups from cost or activity centers may be assigned. When determining appropriate cost components, utility management should consider the utility's budgeting and accounting chart of accounts, types of customers served, and services provided. Cost components should reflect the most significant criteria of facility design,

TABLE 6.1 Example of Wastewater Cost or Activity Center

Operation and maintenance	Facility investment related
Collection	Intangible facility
Pumping	Collection facility
Treatment and disposal	System pumping facility
Customer accounts	Treatment and disposal facility
Administrative and general	Water resource recovery facility
Reclaimed water treatment	Reclaimed water distribution facility
Reclaimed water distribution	General facility
Pretreatment	

operations, and services provided that correspond to distinct differences in customer classes' service requirements.

Typical cost components consist of annual average daily volume, peak flowrates, raw influent wastewater strengths, and customer categories as may typically apply to retail service. Based on the nature of the customer base and service provided, some additional cost-component refinements (e.g., direct, common-to-all, specific-use services, seasonal variations, and growth-related components) may need to be identified and used in conjunction with the typical cost components cited previously. Table 6.2 summarizes the various typical and refining components.

Examples of direct assignment components include pretreatment and strengths-monitoring programs, customer meters and connections, billing and collection, and reclaimed water. Common-to-all-classes and specific service-use components help separate costs for wholesale and retail customers. Seasonal variations could distinguish capacity in system facilities that serve year-round customers from capacity dedicated to serve peak seasonal use customers. Separation of annual costs between existing and growth customers helps determine system development charges that recover growth-related annual operating and capital costs from new connections should this approach reflect the utility's cost recovery and rate policy.

6.0 COST-CAUSATIVE FACTORS

Within each of the utility costs or activity centers discussed previously, one or more of the representative cost-causative factors, shown in Table 6.3, may apply. The cost-causative factor for a particular system component or

TABLE 6.2 Example of Wastewater Cost Components

Typical components	Refining components*
Annual average daily volume	Direct cost component assignment
Peak flow or capacity	Common-to-all-customers
Biochemical oxygen demand	Common-to-a-specific service use
Total suspended solids	Seasonal peak use
Total Kjeldahl nitrogen and/or phosphorus	Year-round use
Pretreatment	Growth-related assignment
Customer-related	Existing customer assignment
Reclaimed water	

*Services-related refinements for cost components.

TABLE 6.3 Example of Representative Wastewater Utility Cost-Causative Factors

Infiltration/inflow volumes	Extraneous water flows that enter the wastewater system from groundwater seeping into buried pipes through cracks or loose joints or from stormwater entering the system through manhole covers, open construction drains, legal or illegal downspouts, and foundation drain connections.
Number of customers	Typically, customers are measured by the number of accounts, with one meter and service connection per account, or by the number of bills issued annually to recover annual customer-related costs. Typically, meters by various sizes and classes are totalled by some type of meter equivalency factor when compared to the smallest meter size in the system. Meter equivalents recover costs associated with measuring wastewater volumes used for billing purposes and costs for maintaining various-sized connections to the system's sewers. With computer billing, cost for billing and collecting customer bills is about the same regardless of meter size or contributed wastewater volumes.
Wastewater volume	Typically, contributions of wastewater by customers are measured by water meters serving the property, often with adjustments for consumptive uses of water, or by sewer meters recording discharges to the system from the property of larger customers. Wastewater and I/I volumes are used as design criteria for wastewater system facilities.
Peak wastewater volume or capacity	Wastewater utilities are sized for combined customer- and weather-related peak flows. The timing and causes of peak flows vary significantly across the United States. Various customer classes' peak wastewater flows occur at different times of the year. Typically, residential customers peak on weekday mornings when preparing for the workday and may peak in the winter when faucets are turned on to keep pipes from freezing. Industrial customers tend to provide peak volumes during the shutdown and cleanup of a process run. Commercial customers tend to peak seasonally, for example, as a result of a significant influx of convention visitors to the community.

(*continued*)

TABLE 6.3 Example of Representative Wastewater Utility Cost-Causative Factors (*Continued*)

Wastewater strength	Before starting the allocation, the influent strength received, the parameters used to operate the facility, and regulatory requirements should be considered. Strengths include soluble and insoluble organic and inorganic matter delivered to the system by customers or as a result of I/I that needs to be removed or neutralized to specified levels during the treatment process for the utility to meet state and federal discharge permit requirements. Water resource recovery facilities are sized using design criteria for BOD, chemical oxygen demand (COD), total Kjeldahl nitrogen, phosphorus, and total TSS.
Biosolids quantity produced	As a residual byproduct of the treatment process, sludge must be further treated before it may be hauled and deposited off-site or used for other purposes. The particular forms of processing and hauling are reflected in facility design and annual operating costs.
Kilometers (miles) of mains by size	The extensiveness of the utility system's service area affects annual capital and operating costs. Typically, mains less than 30.5 cm (12 in.) are used as local collector mains to transport wastewater from customers' property to interceptor mains of 30.5 cm and larger that carry the wastewater to WRRFs. The demarcation between collector and interceptor mains in terms of kilometers (miles) of main by size helps allocate main costs between backbone and local service facilities.
Collection system acreage	The size of the system has a direct bearing on the lengths of mains required to serve customers and, in turn, on collection system capital costs and annual operating costs. Without consideration of density, customer classes occupying the most acreage in the service area should proportionally be responsible for the most collection system costs.
Collection system density	When customers within classes are relatively close together, the investment in collection system costs are less than for those customers who are relatively spread out. Thus, density, or the ability to serve more customers per length of pipe, may have a significant effect on the level of collection system operating and capital costs.

activity is defined as that measurable design criterion, operational or service purpose, or customer requirement that *predominantly* influences the size and annual operating and capital costs of the utility facility or cost center being considered. For example, the causative factor for the capital costs of a large sewer is the design peak flowrate typically. In some cases, multiple cost-causative factors must be considered to form the total cost determinant for a particular facility or cost center. For a WRRF, operating expense and capital cost-causing factors could include annual average daily flow, capacity or peak flowrate, and two or more strength parameters. For the utility's billing and collection function, the cost-causative factors could include the number and relative size of customers served and the number of bills rendered.

In a practical sense, the use of cost-causative factors is limited to those for which information are available or that can be determined or estimated with reasonable effort. Determination of the potential for use of separate cost-causative factors for capital costs and operating expenses involves engineering judgment and knowledge of system operations. Care should be taken to ensure that all aspects are properly considered and that there is a reasonable basis for selected cost-causative factors. Also, fairness in cost recovery requires consistent use of cost-causative factors within each study and over time for each wastewater system, unless some significant operational change warrants a change in applicable factors.

7.0 FUNDAMENTAL COST-ALLOCATION APPROACHES

The wastewater utility industry uses two basic cost-allocation approaches. Each approach recovers operating expenses and capital-related costs, also known as *net revenue requirements*, but how processes are allocated differs, as illustrated in Table 6.4.

The *design-basis cost-allocation methodology* allocates costs based on the design of system components. For example, a secondary basin is designed to meet flow requirements, the aeration equipment would be designed to

TABLE 6.4 Illustration of Difference Between Design- and Functional-Basis Cost-Allocation Methodologies

System/process	Design basis	Functional basis
Secondary basin	Flow	BOD
Aeration	BOD	BOD
Primary basin	Flow	TSS
Solids pumping	TSS	BOD, TSS

meet BOD requirements, the primary basin is designed based on flow, and solids pumping is designed based on the amount of TSS. Proponents of this approach state that utilities predominately incur capital cost and operating expenses and provide service to customers generally on the same basis as they are designed (see Table 6.5).

The *functional-basis cost-allocation methodology* allocates costs based on the function each process serves. For example, a secondary basin functions to remove BOD, so its costs would be substantially allocated to BOD and aeration basins would also be allocated to BOD; conversely, the primary basin functions to remove TSS, so these costs would be allocated to the TSS function like solids pumping. Proponents of this approach state that the function of various cost centers or activities should be the basis for allocating costs, and that this may be easier for the public to understand than the design basis (see Table 6.8).

Typically, the methodology chosen depends on the information available. The aforementioned approaches tend to use different cost-causative factors to allocate costs to cost components. The basic processes, fundamental differences, and advantages and limitations of the two methodologies will be discussed in the remainder of this chapter.

Although not covered in this manual, some utilities use hybrid cost-allocation approaches that mix the design and functional bases of the allocation of costs to cost components. In such hybrid approaches, typically, the design-basis methodology is used to allocate capital costs that reflect the design criteria's significant effect on sizing and construction costs. The functional cost basis is applied to operating expenses that may be more influenced by variations in actual or current operating results. The resulting effects on cost allocations and distribution of costs to customer classes place the results of hybrid methods somewhere between those for the design- and functional-based cost allocations, depending on how cost-causative factors are used.

8.0 DESIGN-BASIS COST-ALLOCATION METHODOLOGY

The design-basis methodology uses the utility's investment in fixed assets to provide a pattern for assigning operating expenses and capital costs to cost components. Table 6.5 shows utility facility examples designed for the specific functions of collection, transport, treatment, and disposal of wastewater and solids. Under the design-basis methodology, it is assumed that operating expenses are allocable to cost components on the same basis as the allocation of facility investment in specific facilities. The rationale is that the size and related investment in a facility is because of the design loading and

TABLE 6.5 Example of Representative Wastewater Utility Cost-Causative Factors in Design-Basis Cost-Allocation Methodology

Facility and function	Predominant design criteria cost-causative factors
Collection sewers	Peak flowrates determine the size of mains, so costs are assigned to the "capacity" cost component.
Lift and pumping stations	Peak flowrates determine the size of stations, so costs are assigned to the "capacity" cost component.
Interceptor sewers	Peak flowrates determine the size of mains, so costs are assigned to the "capacity" cost component.
Grit or screen chamber	Peak flowrates determine the size of chambers, so costs are assigned to the "capacity" cost component.
Primary settling basin	Settling detention times are based on design average flow, so costs are assigned to the "volume" cost component. If known, costs for equipment used to remove sediment may be assigned to the "TSS" cost component.
Trickling filter basins	Biochemical oxygen demand loading determines the size of the basin, so costs are assigned to the "BOD" strength cost component.
Aeration basins	Aeration basins may be used for BOD as well as nutrient removal. Therefore, it could be assigned to one or more strength components (BOD, TSS, total Kjeldahl nitrogen [TKN], phosphorus). Biochemical oxygen demand loading determines the size of the basin, so costs are assigned to the "BOD" strength cost component. Typically, nitrogen loading determines the size of the basin, so costs are assigned to the "TKN" cost component. Phosphorus is concurrently removed by chemical treatment. If phosphorus determines basin size, cost would be assigned to the "phosphorus" cost component.
Secondary settling basins	Settling detention times are based on design average flow, so costs are assigned to the "volume" cost component. If known, costs for mechanical equipment used to remove sediment may be separately assigned to the "BOD" cost component.

(continued)

TABLE 6.5 Example of Representative Wastewater Utility Cost-Causative Factors in Design-Basis Cost-Allocation Methodology (*Continued*)

Facility and function	Predominant design criteria cost-causative factors
In-facility pipes and pumping	Costs for in-facility pipes and pumping are assigned to the cost component pertaining to their design function: "volume", "TSS", or "BOD".
Disinfection	Peak flowrates determine the size of disinfection facilities, so costs are assigned to the "capacity" cost component.
Solids processing	The total quantity of solids that the facility is designed to handle determines the size of facilities, so costs should be assigned to "TSS", "BOD", and "TKN" cost components in proportion to their respective total facility-design loadings.
Solids disposal	The total quantity of solids that the facility is designed to handle determines the nature of facilities and equipment used for disposal, so costs would be allocated to cost components in the same way as solids processing.
Odor control	Odor control may be viewed as not providing wastewater treatment, but rather providing a community benefit, so costs are assigned to the "customer" cost component. Alternatively, in light of regulatory requirements related to odor control, these costs may be assigned to "TSS", "BOD", and "TKN" cost components in proportion to their respective total facility-design loadings.
Outfall sewer	Peak flowrates determine the size and length of the sewer extending into receiving waterways, so costs are assigned to the "capacity" cost component.
Water resource recovery	Peak flowrates determine the size of filter basins, so costs are assigned to the "capacity" cost component of reclaimed water service.
Reclaimed water distribution	Peak flowrates determine the size of mains, so costs are assigned to the "capacity" cost component of reclaimed water service.

operating expenses associated with each facility are, in turn, proportional to the size of the facility. The design basis works equally well whether utility costs of service are developed using the cash or the utility basis shown in Appendix A. Tables 6.6 through 6.10 use the cash-basis of annual cost-of-service determination. This chapter uses a hypothetical secondary treatment facility with chlorine disinfection for the example tables.

When allocating capital-related revenue requirements, two competing methods, original cost less depreciation (OCLD) and replacement cost less depreciation (RCLD), may be used to assess value. Unless a system is brand new, OCLD is a less ideal way to value asset return because it does not reflect the value of service being received. The use of RCLD would require an estimation of the replacement cost. A mix of both OCLD and RCLD may be used as well.

Given that the design-basis cost-allocation methodology and the utility basis of cost of service both depend on the net facility investment, having a good asset management program may provide a better indication of net facility investment. Typically, depreciation in cost of service has been calculated using standard depreciation methods such as straight-line depreciation. An asset management program provides the opportunity to use an actual rate of deterioration rather than straight-line depreciation or another standard depreciation accounting schedule. Additionally, a good asset management program provides the potential for more accurately projecting the cost of maintenance for specific assets, and/or common-to-all assets, or wholesale-only assets when determining revenue requirements. This is an area to keep in mind, particularly as asset management programs mature and include systematic and consistent condition evaluations.

8.1 Design-Basis Allocation of Facility Investment and Payments in Lieu of Taxes

Table 6.6 provides an example of depreciated original-cost facility investment allocated to cost components that recognizes the predominant design criteria by facility, shown in Table 6.5. As mentioned in Section 1.1, examples reflect generally accepted processes and a utility should tailor its cost allocation to its system. Cost components consist of volume, capacity, TSS, BOD, total Kjeldahl nitrogen (TKN), customer billing, and customer. Allocations unique to this example are as follows:

- Sludge processing and disposal facility total costs in Table 6.6 are allocated 45% to TSS, 40% to BOD, and 15% to TKN cost components based on facility design criteria. Each facility's design should be analyzed to determine the proper percentage split.

TABLE 6.6 Example of Allocation of Net Facility Investment and Capital Costs to Cost Components in Design-Basis Cost-Allocation Methodology

				Wastewater strength				
	Total	Volume	Capacity	TSS	BOD	TKN	Customer billing	Customer
Facility in service								
Influent pumping	$7,558,000.00		$7,558,000.00					
Preliminary treatment	$5,434,000.00		$5,434,000.00					
Primary sedimentation	$3,327,000.00	$3,327,000.00						
Aeration	$11,435,000.00				$11,435,000.00			
Nutrient removal	$2,082,000.00					$2,082,000.00		
Secondary sedimentation	$9,318,000.00	$9,318,000.00						
Chlorination	$1,690,000.00		$1,690,000.00					
Solids handling and treatment	$19,049,000.00			$8,572,000.00	$7,620,000.00	$2,857,000.00		
Dewatering and disposal	$28,788,000.00			$12,955,000.00	$11,515,000.00	$4,318,000.00		
Outfall diversion sewer	$1,101,000.00		$1,101,000.00					
General facility	$3,751,000.00	$528,000.00	$659,000.00	$899,000.00	$1,278,000.00	$387,000.00		
Total facility	$93,533,000.00	$13,173,000.00	$16,442,000.00	$22,426,000.00	$31,848,000.00	$9,644,000.00		
Pumping and lift stations	$4,092,000.00		$4,092,000.00					
Collection system	$79,723,000.00		$71,751,000.00					$7,972,000.00
General facility	$3,986,000.00	$222,000.00	$1,555,000.00	$378,000.00	$537,000.00	$163,000.00	$997,000.00	$134,000.00
Total facility investment	$181,334,000.00	$13,395,000.00	$93,840,000.00	$22,804,000.00	$32,385,000.00	$9,807,000.00	$997,000.00	$8,106,000.00
Less contributed facilities	$49,880,000.00		$45,097,000.00					$4,783,000.00
Total net investment	$131,454,000.00	$13,395,000.00	$48,743,000.00	$22,804,000.00	$32,385,000.00	$9,807,000.00	$997,000.00	$3,323,000.00
Net capital costs	$9,610,000.00	$979,000.00	$3,563,000.00	$1,667,000.00	$2,368,000.00	$717,000.00	$73,000.00	$243,000.00
Payments in lieu of taxes	$975,000.00	$103,000.00	$374,000.00	$175,000.00	$248,000.00	$75,000.00		
Total capital costs	$10,585,000.00	$1,082,000.00	$3,937,000.00	$1,842,000.00	$2,616,000.00	$792,000.00	$73,000.00	$243,000.00

- Administration and general WRRF costs are allocated using an indirect allocation in which the allocation to cost components is in proportion to the total WRRF cost allocations.
- Collection system costs are allocated to capacity and customer cost components. The 10% of collection system costs assigned to the customer cost component reflects an allowance for the relative size of the system and the low density of development's effect on the collection system's total costs. In addition, the utility is assumed to use a minimum collection main size for cleaning purposes. Such sizing is independent of the volume of wastewater carried by the mains.
- General facility item costs (administrative offices, shops, vehicles, and work equipment) provide support to all system facilities. General facility costs, not including customer billing equipment, are allocated to all cost components in proportion to their totals of system original cost allocations. Additionally, in this example, customer billing equipment is assumed to be 25% of total general facility costs and is assigned to the customer billing component.
- Contributed facilities may cover federal and state grants plus developer and individual customer contributions. These will be specific for each agency. For this example, contribution costs are assumed to be limited to collection system small mains (60% of total wastewater collection and 50% of total pumping and lift stations).
- The total net investment is the total investment less contributions.

The resulting facility allocation in Table 6.6 provides a pattern for next allocating cash-basis cost-center revenue requirements of net capital costs, payments in lieu of taxes (PILOT), and certain line-item O&M expenses.

8.1.1 Net Capital Costs

Net capital costs in Table 6.6, consist of annual debt service payments, equipment purchases, and capital improvements made from user-charge revenue, as shown previously in Table 5.3. Although not identified by this example case study, miscellaneous funding by system development charges, interest earnings on invested capital funds, and other capital fees should be deducted from capital costs to be recovered through user rates. Miscellaneous capital funding sources represent revenue that directly offsets capital costs. Net capital costs are allocated to cost components in the same manner as "total net investment" in utility assets. This allocation approach recognizes the following:

- The investment in system assets may reasonably represent the level of investment anticipated for the planning period that new rates are intended to be in place and

- If no new debt issues are anticipated beyond those recognized by revenue requirements development and annual capital expenditures are reasonably uniform regardless of the varying types of expenditure needs during the rate period, then cost recovery from customer classes would remain reasonably consistent for the rate use period.

If cost recovery is reasonably consistent, rates will be reasonably consistent without considerable swings in rate levels during the rate period.

8.1.2 Payments in Lieu of Taxes

In Table 6.6, PILOT is apportioned to the cost components based on total original cost less depreciation of aboveground facility investment. The allocation of PILOT should follow the basis on which the payment is levied, that is, on asset value or gross receipts. If it is based on asset value, as ad valorem taxes are, PILOT should be allocated to cost components in the same manner as aboveground facilities. A separate summary of facility value would need to be developed using only aboveground assets. The allocation of the PILOT cost is based on the value of aboveground assets where aboveground facilities are assumed to include all assets except pipes, 50% of lift and pumping stations, and general facility assets. If the payment is based on gross receipts, then PILOT should be allocated to customers based on class revenue generated.

8.2 Design-Basis Allocation of Net Operation and Maintenance Expenses

Net O&M expenses, by example, are allocated to cost components in Table 6.7 in a similar manner as the respective elements of facility investment in Table 6.6, with the exceptions discussed below. Cost components consist of volume, capacity, TSS, BOD, TKN, customer billing, customer, and industrial surcharge. Typically, design-based facilities are sized larger than current use and provide flexibility for meeting growth while covering current peak user and wet weather demands. Projected O&M expenses reduced by miscellaneous fee revenue and capital funding sources for capitalized O&M expenditures result in the net O&M expenses to be generated by user charges. As mentioned in Section 1.1, examples reflect relatively common processes and generally accepted allocations, but a utility should tailor its cost allocation to its system.

To enhance the assignment of annual costs to cost components, utility management should review utility operations for costs that may be uniquely assigned within their utility. Some itemized costs may be better allocated using a different cost component than one generally applied to all other

TABLE 6.7 Example of Allocation of O&M Expenses to Cost Components in Design-Basis Cost-Allocation Methodology

	Total	Volume	Capacity	Wastewater strength			Customer billing	Customer	Industrial surcharge
				TSS	BOD	TKN			
Water resource recovery facilities									
Influent pumping	$636,000.00	$223,000.00	$413,000.00						
Preliminary treatment	$357,000.00		$357,000.00						
Primary sedimentation	$190,000.00	$190,000.00							
Aeration	$968,000.00				$968,000.00				
Nutrient removal	$450,000.00					$450,000.00			
Secondary sedimentation	$319,000.00	$319,000.00							
Chlorination	$393,000.00	$138,000.00	$255,000.00						
Solids handling and treatment	$1,126,000.00			$507,000.00	$450,000.00	$169,000.00			
Dewatering and disposal	$2,783,000.00			$1,252,000.00	$1,113,000.00	$418,000.00			
Laboratory	$410,000.00			$185,000.00	$164,000.00	$61,000.00			
Subtotal treatment	$7,632,000.00	$870,000.00	$1,025,000.00	$1,944,000.00	$2,695,000.00	$1,098,000.00			
Wastewater transmission									
Pumping and lift stations	$503,000.00	$176,000.00	$327,000.00						
Equipment maintenance	$770,000.00		$770,000.00						
Supervision	$456,000.00		$456,000.00						
Subtotal transmission	$1,729,000.00	$176,000.00	$1,553,000.00						
Wastewater collection	$3,667,000.00		$3,300,000.00					$367,000.00	
Engineering	$795,000.00	$30,000.00	$344,000.00	$114,000.00	$158,000.00	$64,000.00		$53,000.00	$32,000.00
Industrial waste administration	$1,093,000.00							$547,000.00	$546,000.00
Billing and collection	$1,142,000.00						$1,142,000.00		
All other	$640,000.00	$43,000.00	$248,000.00	$82,000.00	$114,000.00	$46,000.00	$46,000.00	$39,000.00	$23,000.00
Office of the director	$502,000.00	$34,000.00	$195,000.00	$64,000.00	$89,000.00	$36,000.00	$36,000.00	$30,000.00	$18,000.00
Planning and project management	$1,002,000.00	$67,000.00	$388,000.00	$128,000.00	$178,000.00	$73,000.00	$71,000.00	$60,000.00	$36,000.00
Total O&M	$18,202,000.00	$1,220,000.00	$7,053,000.00	$2,332,000.00	$3,234,000.00	$1,317,000.00	$1,295,000.00	$1,096,000.00	$655,000.00
Operation and maintenance reserve	$66,000.00	$4,000.00	$26,000.00	$8,000.00	$12,000.00	$5,000.00	$5,000.00	$4,000.00	$2,000.00
Less revenue from other sources	$17,000.00	$1,000.00	$7,000.00	$2,000.00	$3,000.00	$1,000.00	$1,000.00	$1,000.00	$1,000.00
Net O&M expense	$18,251,000.00	$1,223,000.00	$7,072,000.00	$2,338,000.00	$3,243,000.00	$1,321,000.00	$1,299,000.00	$1,099,000.00	$656,000.00

costs of the same cost center. Some costs may be assigned directly to a specific cost component because of unique service that the utility provides to a limited number of customers. Other costs may have general applicability to all cost components and may be proportioned to all components based on individual totals of all other allocated component costs. The following are some examples:

- Some annual consumables' costs for WRRF components designed to meet capacity requirements may be more dependent on annual wastewater volume than peak flowrates. Such is the case for pumping, solids settling, chlorination, phosphorus removal, and UV disinfection facilities. Purchased power and chemical costs should be allocated to the volume cost component to reflect that power consumed and design rates of chemical feed for treatment processes will be consistently used throughout the year for the total flow handled by the facility. The electric rate schedule used by the utility may have a volume and a demand component. Should this be the case, the demand charge component of the bill should be allocated to the capacity cost component.
- For a reasonably homogeneous and compact service area, O&M expenses for local collector sewers are allocated to the capacity cost component typically. They could also be allocated to the customer cost component based on density of development and the extensiveness of the collection system. For this example, 10% of collection system costs are allocated to the customer cost component to recognize a low density of development within the system, and the remainder of the costs are allocated to the capacity component.
- Some costs should be directly assigned to a particular cost component. If the wastewater utility has an established industrial monitoring and surveillance program, there may be significant industrial waste costs that should be directly assigned to either pretreatment or industrial surcharge aspects of the program. Additionally, a proportional share of laboratory costs may be appropriately allocated to these programs. Customer billing and collection costs should be directly assigned to the customer cost component because these systems are designed for the number of customers served, not for contributed strengths and volumes of wastewater.
- Engineering, planning, and project management should be allocated to cost components proportionally to typical or historical operational support time provided by personnel working in these areas; alternatively, they may be allocated in proportion to net assets (see Table 6.7).

- General and administrative expenses are allocated to cost components in proportion to the total of all other expenses, less costs for chemicals and power. An alternative approach may be to allocate general and administrative expenses entirely to the customer cost components. This approach would require some level of examination of the specific costs included in general and administrative expenses. For example, if costs are primarily related to personnel-related costs like employee benefits, an allocation based on labor costs may be appropriate. Conversely, if costs are known to be exclusively related to nonpersonnel administrative overhead items, it may be reasonable to allocate these costs entirely to the customer component.

8.3 Additional Design-Basis Cost-Allocation Considerations

The following are additional design-basis cost-allocation considerations:

- Costs for WRRF safety and security should be allocated in proportion to the total costs for other cost parameters, much like general facility and administrative and general operating costs;
- Costs for odor and vector control may be allocated to the customer cost component for the affected WRRF or collection system facilities or assigned to "TSS", "BOD", and "TKN" cost components in proportion to their respective total facility-design loadings; and
- Costs for corrosion control should be assigned to the functional cost component using the design basis for the affected collection system or WRRF component.

8.3.1 Approach Advantages

The design-basis cost-allocation methodology, by tying together the design of a utility's fixed assets with operating expenses and customer class use patterns (through the use of the same cost components), provides a practical means of demonstrating how a utility incurs its costs and what it costs a utility to serve each class. The allocation process is logical and follows documented engineering design criteria. Furthermore,

- Design criteria from engineering reports and limits from regulatory permits for wastewater utilities provide documentation of the defensibility of design-basis methodology results.
- Operating expenses and capital costs may be allocated to cost components using the allocation of facility investment as a pattern, thereby

recognizing that the utility incurs costs for both currently used and extra capacity in facilities that are operated and maintained annually.

- The design-basis approach provides a means of recognizing system design and class capacity and strengths requirements, as well as contributed total annual volumes.
- The design basis readily demonstrates how the utility fairly and equitably recovers operation, maintenance, and replacement (OM&R) capital costs from customers served.
- For utilities with wholesale wastewater service, the contract for such service often stipulates a contractual capacity for the annual average daily rate of flow. Contract terms may also include a contractual peak flowrate and contractual strength parameters for the wholesale customer. The design basis of cost allocation provides a direct linkage for allocating investment- and capital-related costs to wholesale service customers based on the design capacity of the various facilities used by the wholesale customer and the contractual capacities set forth in the wholesale contract.

8.3.2 Approach Limitations

Some utilities do not maintain records of their fixed assets in sufficient detail by utility function to readily use the design-basis methodology. In such cases, an engineer's surrogate estimates of asset design and cost details based on facilities inspections and as-built drawings may be required. Some utilities without a requirement to use monitoring or pretreatment programs may not maintain sufficient records by class of service to permit differentiation by volume and strength cost-causative parameters. Some utilities' billing records are not maintained by typical utility classes of service covering residential, commercial, institutional, and industrial customers, making allocations to classes difficult, if not impossible.

9.0 FUNCTIONAL-BASIS COST-ALLOCATION METHODOLOGY

Another frequently used cost-allocation approach is the functional-basis cost-allocation method. Costs by cost center or activity center are allocated to cost components based on the purpose (or function) of the cost center/activity. Some purpose-based cost-causative factors are shown in Table 6.8.

The functional-basis cost-allocation method works well with governmental and utility-based charts of accounts. When utility charts of accounts are used, total costs by utility cost or activity centers are readily identifiable. When governmental charts of accounts are used, additional effort is required

TABLE 6.8 Example of Representative Wastewater Utility Cost-Causative Factors in Functional-Basis Cost-Allocation Methodology

Facility and function	Predominant purpose cost-causative factors
Collection system	The sewers' purpose is to carry wastewater at variable flowrates, so costs are assigned to the "volume" cost component. In the case of hybrid approaches, these costs may be assigned to both a volume and capacity component.
Lift and pumping stations	The stations' purpose is to move wastewater at variable flowrates, so costs are assigned to the "volume" cost component. In the case of hybrid approaches, these costs may be assigned to both a volume and capacity component.
Interceptor sewers	The sewer' purpose is to carry wastewater at variable flowrates, so costs are assigned to the "volume" cost component. In the case of hybrid approaches, these costs may be assigned to both a volume and capacity component.
Grit or screen chamber	The chamber's purpose is to remove grit, debris, and larger material carried by wastewater. While some of this material is not included in the analysis of "TSS", grit and other smaller items are, so costs are assigned to the "TSS" cost component typically.
Primary settling basin	The settling detention basin's purpose is to remove TSS, so costs are assigned to the "TSS" cost component.
Trickling filter basins	The basins' purpose is to remove BOD from wastewater, so costs are assigned to the "BOD" cost component.
Aeration basins	The basins' purpose is to remove BOD from wastewater, so costs are assigned to the "BOD" cost component. If the facilities' purpose is nitrogen removal, costs are assigned to the "TKN" cost component. Phosphorus is concurrently removed by chemical treatment, so such chemical costs are assigned to the "volume" cost component. If the facilities' purpose is phosphorus removal, cost would be assigned to the "phosphorus" cost component. If the facilities' purpose is to address more than one XXX, then it can be split based on YYY.
Secondary settling basin	The basins' purpose is to remove BOD from wastewater, so costs are assigned to the "BOD" cost component.

(continued)

TABLE 6.8 Example of Representative Wastewater Utility Cost-Causative Factors in Functional-Basis Cost-Allocation Methodology (*Continued*)

Facility and function	Predominant purpose cost-causative factors
In-facility pipes and pumping	Costs for in-facility pipes and pumping are assigned to the cost component pertaining to their purpose: "volume", "TSS", or "BOD".
Disinfection	The facilities' purpose is to reduce bacteria and other pathogens at variable flowrates, so costs are assigned to the "volume" cost component.
Solids processing	The facilities' purpose is to handle and process solids, so costs should be assigned to "TSS", "BOD", and "TKN" cost components in proportion to their respective current annual loadings at the facility.
Solids disposal	The facilities' purpose is to dispose of solids, so costs should be assigned to "TSS" and "BOD" cost components in proportion to their respective current annual loading at the facility.
Odor control	Odor control's purpose is to contain objectionable smells at the WRRF or specific system facilities, so costs are assigned to the cost component representative of the purpose for each facility requiring odor control items.
Outfall sewer	The sewers' purpose is to carry wastewater at variable flowrates, so costs are assigned to the "volume" cost component. In the case of hybrid approaches, these costs may be assigned to both a volume and capacity component.
Water resource recovery	The facility's purpose is to further treat wastewater at variable flowrates, so costs are assigned to the "volume" cost component of reclaimed water service. If costs associated with reclaimed water treatment are recovered through reclaimed water rates, they should be excluded for the purposes of calculating sewer service rates. See Chapter 13 for details on calculating reclaimed water rates.
Reclaimed water distribution	The purpose is to carry wastewater at variable flowrates, so costs are assigned to the "volume" cost component of reclaimed water service. In the event that costs associated with reclaimed water distribution are recovered through reclaimed water rates, they should be excluded for the purposes of calculating sewer service rates. See Chapter 13 for details on calculating reclaimed water rates.

to assign revenue requirements to cost or activity centers. Interviews with knowledgeable utility staff provide estimates of the percentage of annual personnel time spent in each utility cost center. Reviews of electric bills, chemical invoices, and other material and supply purchases with knowledgeable utility staff provide indications of which bills pertain to particular cost centers. Either approach results in cost-of-service revenue requirements by cost centers that may be readily allocated to cost components by the purpose-based cost-causative factors shown in Table 6.8.

9.1 Functional-Basis Allocation of Net Facility Investment and Payments in Lieu of Taxes

Table 6.9 provides an example of depreciated original-cost facility investment allocated to cost components, which recognizes the predominant operating and maintenance cost-causative purposes by facility shown in Table 6.8. As mentioned in Section 1.1, examples reflect generally accepted processes, and a utility should tailor its cost allocation to its system. Cost components consist of volume, TSS, BOD, TKN, customer billing, and customer. Capital costs in of Table 6.9 reflect determination of cash-basis revenue requirements. As with the design basis, when allocating capital-related revenue requirements, OCLD and RCLD may be used to assess value. Allocations unique to this example are as follows:

- In this example, solids processing and disposal facility costs are allocated 42% to TSS, 45% to BOD, and 13% to TKN functional cost components, representing the current levels of annual loadings. However, each facility's operating records should be accessed to determine actual loadings on which to base percentage splits.
- Collection system costs are allocated to volume and customer cost functions. In this example, the cost of facility investment is assumed to be allocated 90% to flow and 10% to customer. The percent split for the transmission and collection system will depend, in part, on how much investment is made in replacement of small mains.
- An assumed 25% of general facility costs covers the wastewater utility's share of customer billing equipment.
- Contributed facilities may cover federal and state grants plus developer and individual customer contributions. For this example, contribution costs are assumed to be limited to collection system small mains and lift stations. For this example, contribution costs are assumed to be limited to collection system small mains (60% of total wastewater collection and 50% of total pumping and lift stations).
- The net investment is the total investment less contributions.

TABLE 6.9 Example of Allocation of Net Facility Investment and Capital Costs to Cost Components in Functional-Basis Cost-Allocation Methodology

			Wastewater strength				
	Total	Volume	TSS	BOD	TKN	Customer billing	Customer
Water resource recovery facility in service							
Influent pumping	$7,558,000.00	$7,558,000.00					
Preliminary treatment	$5,434,000.00		$5,434,000.00				
Primary sedimentation	$3,327,000.00		$3,327,000.00				
Aeration	$11,435,000.00			$11,435,000.00			
Nutrient removal	$2,082,000.00				$2,082,000.00		
Secondary sedimentation	$9,318,000.00			$9,318,000.00			
Chlorination	$1,690,000.00	$1,690,000.00					
Solids handling and treatment	$19,049,000.00		$8,001,000.00	$8,572,000.00	$2,476,000.00		
Dewatering and disposal	$28,788,000.00		$12,091,000.00	$12,955,000.00	$3,742,000.00		
Outfall diversion sewer	$1,101,000.00	$1,101,000.00					
General facility	$3,751,000.00	$432,000.00	$1,205,000.00	$1,766,000.00	$348,000.00		
Total facility	$93,533,000.00	$10,781,000.00	$30,058,000.00	$44,046,000.00	$8,648,000.00		
Pumping and lift stations	$4,092,000.00	$4,092,000.00					
Collection system	$79,723,000.00	$71,750,700.00					$7,972,300.00
General facility	$3,986,000.00	$1,460,000.00	$507,000.00	$742,000.00	$146,000.00	$997,000.00	$134,000.00
Total facility investment	$181,334,000.00	$88,083,700.00	$30,565,000.00	$44,788,000.00	$8,794,000.00	$997,000.00	$8,106,300.00
Less contributed facilities	$49,880,000.00	$45,097,000.00					$4,783,000.00
Total net investment	$131,454,000.00	$42,986,700.00	$30,565,000.00	$44,788,000.00	$8,794,000.00	$997,000.00	$3,323,300.00
Net capital costs	$9,610,000.00	$3,143,000.00	$2,234,000.00	$3,274,000.00	$643,000.00	$73,000.00	$243,000.00
Payments in lieu of taxes	$975,000.00	$330,000.00	$234,000.00	$344,000.00	$67,000.00		
Total capital costs	$10,585,000.00	$3,473,000.00	$2,468,000.00	$3,618,000.00	$710,000.00	$73,000.00	$243,000.00

9.1.1 Net Capital Costs

Net capital costs consist of annual debt service payments, equipment purchases, and capital improvements funded with user-charge revenue, as shown previously in Table 5.3. Although not identified by this example case study, miscellaneous funding by system development charges, interest earnings on invested capital funds, and other capital fees should be deducted from capital costs to be recovered through user rates. Miscellaneous capital funding sources represent revenue that directly offsets capital costs. Net capital costs are allocated to functional cost components in the same manner as total net investment in utility assets. This allocation approach recognizes the following:

- The investment in system assets reasonably represents the level of investment anticipated for the planning period that new rates are intended to be in place;
- If no new debt issues are anticipated beyond those recognized by revenue requirements development and annual capital expenditures are reasonably uniform regardless of the varying types of expenditure needs during the rate period, cost recovery from customer classes would remain reasonably consistent for the rate period; and
- If cost recovery is reasonably consistent, rates will be reasonably consistent without considerable swings in rate levels during the rate period.

9.1.2 Payments in Lieu of Taxes

For this example, PILOT is shown in Table 6.9, is assumed to be based on current asset value for aboveground facilities, and is computed in the same manner as ad valorem taxes are computed for similar investor-owned utility property. The assessor's office should be able to identify or estimate the current appraised value of each aboveground utility facility. Care should be taken by the assessor's office to recognize differences in commercial and utility property costs when determining current value.

9.2 Functional-Basis Allocation of Operation and Maintenance Expenses

Operation and maintenance expenses apply to wastewater treatment, solids disposal, industrial waste pretreatment, wastewater transmission, the collection system, and administration functional cost centers. Cost components consist of volume, TSS, BOD, TKN, customer billing, customer, and industrial surcharge. Net O&M expenses exclude miscellaneous revenue from other sources.

The O&M expenses, by example, are allocated to cost components in Table 6.10 based on the functions served by those O&M activities. These allocations are made based on a combination of direct assignment of costs (e.g., polymer costs are incurred because of treating solids and BOD so those costs are allocated to those cost centers), breakdown of operator and maintenance crew time at the activity centers, and best engineering judgment. Miscellaneous costs within the WRRF are allocated the same as the total of the other WRRF allocations. Items like administration and general facility are allocated the same as all the other line items (treatment, collection system, etc.), reflecting that these expenses are incurred because of managing the entire treatment system. As mentioned in Section 1.1, examples reflect common processes and generally accepted allocations, and a utility should tailor its cost allocation to its system.

Allocations unique to this example are as follows:

- Typically, costs by wastewater treatment and solids removal cost centers are allocated to flow, BOD, TSS, and nitrogen cost components based on estimates of how personnel costs, chemicals, power, solids removal, and other budgeted line-item costs are used during the year. In this example, for personnel and other, the percentages used are 42% for TSS, 45% for BOD, and 13% for TKN, paralleling the current loadings handled by the facility.
- Power costs are split between volume and BOD because a lot of the power costs are attributable to running the aeration mixers.
- Chemical costs are allocated to volume, BOD, and TSS in this example because chemicals are often used to address BOD and TSS. To the extent available, utility records should be used to make the best engineering judgment about allocation percentages. Any costs for odor, vector, and corrosion control would be allocated to cost components using the same purpose bases as those used for the affected WRRF and collection system facilities.
- Costs related to transmission are allocated to volume because the main function of transmission-related costs is to move the flow.
- Collection system costs are allocated to volume and customer cost functions. For the example, 50% of collection system costs are assigned to the customer function, reflecting that, to ease connections and maintenance, collector sewers are set at a minimum size, which is frequently much greater than what is required to carry anticipated flows. Therefore, the cost is assumed to be allocated 50% to flow and 50% to customer in this example.
- Administration costs for "office of director", "planning and project management", and "all other expenses" are assigned on the same

TABLE 6.10 Example of Allocation of O&M Expenses to Cost Components in Functional-Basis Cost-Allocation Methodology

			Wastewater strength					
	Total	Volume	TSS	BOD	TKN	Customer billing	Customer	Industrial surcharge
Wastewater treatment								
Personnel	$1,861,000.00		$782,000.00	$837,000.00	$242,000.00			
Power	$745,000.00	$372,000.00		$373,000.00				
Chemicals	$372,000.00	$124,000.00	$124,000.00	$124,000.00				
Other	$745,000.00	$123,000.00	$227,000.00	$334,000.00	$61,000.00			
Solids disposal								
Personnel	$1,954,000.00		$821,000.00	$879,000.00	$254,000.00			
Power	$782,000.00		$329,000.00	$351,000.00	$102,000.00			
Chemicals	$391,000.00		$164,000.00	$176,000.00	$51,000.00			
Other	$782,000.00		$329,000.00	$351,000.00	$102,000.00			
Industrial waste	$1,093,000.00						$547,000.00	$546,000.00
Subtotal treatment	$8,725,000.00	$619,000.00	$2,776,000.00	$3,425,000.00	$812,000.00		$547,000.00	$546,000.00
Wastewater transmission								
Pumping and lift stations	$503,000.00	$503,000.00						
Equipment maintenance	$770,000.00	$770,000.00						
Supervision	$456,000.00	$456,000.00						
Subtotal transmission	$1,729,000.00	$1,729,000.00						
Wastewater collection	$3,667,000.00	$1,833,000.00					$1,834,000.00	
Administration								
Billing and collection	$1,142,000.00					$1,142,000.00		
Office of the director	$502,000.00	$137,000.00	$91,000.00	$113,000.00	$27,000.00	$38,000.00	$78,000.00	$18,000.00
Engineering	$795,000.00	$260,000.00	$185,000.00	$271,000.00	$53,000.00	$6,000.00	$20,000.00	
Planning and project management	$1,002,000.00	$275,000.00	$182,000.00	$225,000.00	$53,000.00	$75,000.00	$156,000.00	$36,000.00
All other expenses	$640,000.00	$175,000.00	$116,000.00	$144,000.00	$34,000.00	$48,000.00	$100,000.00	$23,000.00
Subtotal administration	$4,081,000.00	$847,000.00	$574,000.00	$753,000.00	$167,000.00	$1,309,000.00	$354,000.00	$77,000.00
Total O&M	$18,202,000.00	$5,028,000.00	$3,350,000.00	$4,178,000.00	$979,000.00	$1,309,000.00	$2,735,000.00	$623,000.00
Operation and maintenance reserve	$66,000.00	$18,000.00	$12,000.00	$15,000.00	$4,000.00	$5,000.00	$10,000.00	$2,000.00
Less revenue from other sources	$17,000.00	$4,000.00	$3,000.00	$4,000.00	$1,000.00	$1,000.00	$3,000.00	$1,000.00
Net O&M expense	$18,251,000.00	$5,042,000.00	$3,359,000.00	$4,189,000.00	$982,000.00	$1,313,000.00	$2,742,000.00	$624,000.00

basis as the total of all other activities. These allocation processes recognize that, while the exact nature of administrative activities during any year cannot be known in advance with certainty, they will generally proportionally support the operations and services represented by the cost components over time, if not each year.

- Administration—engineering costs are allocated per the net facility investment allocations because engineering works are most often related to assets.

9.3 Additional Functional-Basis Cost-Allocation Considerations

The following are additional functional-basis cost-allocation considerations:

- Some utility activities may benefit only specific users, such as waste pretreatment programs for industrial customers and customer billing and collection services. Such costs should be assigned directly to their corresponding cost components.
- Costs for WRRF safety and security should be added in proportion to the total costs for other cost parameters, much like general facility and administrative and general operating costs.
- Certain general fund expenses may be administrative or overhead costs that support multiple departments, with services related to the sewer utility allocated as indirect costs from the general fund to the sewer utility to be recovered through sewer user charges. Unless a utility has specific detail for line-item allocations and a basis for assigning indirect costs to specific cost components, the allocation of indirect costs should be allocated to each cost component in proportion to the overall cost allocations because the costs associated with indirect costs from the general fund support the entire sewer operation typically.

9.3.1 Approach Advantages

Functional-basis cost-allocation methodology, by assigning the revenue requirements by cost or activity centers to cost components based on the function or purpose of the cost center, provides a practical means of demonstrating how a utility incurs its costs and what it costs a utility to serve each class. The allocation process is logical without requiring knowledge of specific design parameters. Furthermore,

- The approach may be used when fixed asset, treatment, and customer billing records are incomplete or when engineering design by system and facility components is not readily known;

- Simplified cost-allocation and rate-development approaches are relatively easy for policymakers and the general public to understand;
- Operating expense and capital costs may be allocated to cost components using the purpose-based allocation of facility investment as a pattern to simplify the allocation process; and
- The functional-basis cost-allocation methodology demonstrates how the utility fairly and equitably recovers OM&R costs from customers served.

9.3.2 Approach Limitations

In its basic form, the functional-basis cost-allocation methodology for allocations of cost or activity centers to cost components is based, typically, on knowledgeable utility staff and rate analysts' estimates and opinions about a particular facility's purpose, rather than engineering design documentation. Although these estimates may reference information sources, such as costs recorded by utility charts of account, detailed facility investment records, and facility engineering design reports, these estimates and opinions are more subjective than relying on specific engineering design information.

Another limitation of the functional-basis cost-allocation methodology is that it may be based on current use and not recognize capacity built into the system to serve future customers or the way that capacity should be recovered from existing customers served. By default, when available or unused capacity is not separately recognized, the methodology assumes that capacity requirements are all the same for all customers and are reflected proportionately by each class's currently contributed volumes. Current-use volume proportionalities may not reflect customer class service capacity requirements used for system design. Large-volume users tend to pay a greater share of total costs than they might otherwise have to bear if capacity is not recognized by the allocation methodology. When significant capacity is unused because the customers for which it was built have not arrived or because customers have left, hybrid approaches may be used to ensure that selected capacity-related costs may be assigned to both volume and capacity components.

As a corollary to the current-use limitation described above, functional-basis cost-allocation methodology also does not lend itself as well as design-basis cost-allocation methodology in the assignment of contract capacity-related costs for wholesale service customers. Typically, a wholesale customer's service contract specifies capacity limits for average daily flow, peak flowrates, and strength parameters. Design-basis cost-allocation methodology is better suited for addressing the appropriate sharing of capital costs based on contract customer capacities.

10.0 WHOLESALE/OUTSIDE CITY CUSTOMER CLASS CONSIDERATIONS

If a utility serves wholesale or outside city customers, the utility basis of cost allocation is often recommended. The utility basis is discussed in detail in Appendix A. However, the basic premise is the parallel between investor-owned utilities serving their customers, who are not owners, and a government-owned utility providing services to customers who are not owners of the system (e.g., outside city users). The government-owned utility is entitled to a reasonable return from non-owner customers based on the value of facility investment required to serve those customers because, if those customers choose to leave the system or reduce their demand, the government-owned utility is left with assets or capacity it no longer needs.

If wholesale or outside city service is involved, more refined cost components are needed. These refinements separate those costs that are common to all customers from common-to-collection or inside city or retail only and from wholesale/outside city only. Typically, the common-to-all heading covers system backbone facilities, and the common-to-collection/inside city-only heading covers direct assignment of small collection mains and lift stations in the owning utility's service area. Inside city/retail-only service includes both common-to-all and retail-only costs. Wholesale service may comprise common-to-all costs and wholesale-only costs. In an instance where a utility may have several WRRFs and wholesale customers are served by one of those facilities, wholesale customers would share in the costs related to that facility but not other WRRFs. Wholesale-only costs may include identifying assets (and their costs) used to serve the wholesale customer and any operating costs directly related to providing that wholesale customer service.

11.0 LEGAL GUIDELINES

When setting rates, utility management must recognize that it should not act in a vacuum and may be called upon to justify the development of the utility's rates and charges and the way they have been applied to customers' bills. Typically, although courts may overturn rates and charges for specific reasons, courts recognize that rate-setting is a legislative function. No generalized court-ordered cost recovery and rate methodology exists. Courts have ordered the implementation of cost-recovery methodologies mutually agreed to by the parties involved in negotiated settlements of lawsuits. Arbitration/mediation has also been used. Courts often remand overturned rates and charges to the governing legislative body for remediation, whether that body is a municipal governing body, utility board, or regulatory commission.

Prudent utility management should seek legal advice regarding utility case law in the state where the utility operates before significantly modifying the utility's historical cost-recovery and rate-setting practice. General findings of the courts pertaining to municipal cost allocations and rate-setting should be considered by management during the development of a utility's schedule of charges for service. A sample of general precepts sometimes referenced in judicial decisions regarding utility rates is as follows:

- Rates established in a lawful manner by a municipality, or municipal agency, are presumed to be reasonable, fair, and lawful;
- Parties challenging municipal rates bear the burden of proving that the rates charged are unjustly discriminatory and unreasonable;
- Once sufficient evidence is introduced that challenges the validity of the presumption that legally set municipal rates are reasonable, fair, and lawful, the presumption is no longer valid;
- Municipal rates must bear some relationship to present or future costs of providing utility service or the courts may deem a rate to be a tax whose sole purpose is generating revenue;
- Rates charged by the municipality must be reasonable and free from unjust discrimination among the customers the municipality serves by taking into account their situation and classification;
- Municipalities may define customer classes and assign similarly situated accounts to the classes based on such factors as differences in the amount of service used, character of service, purpose of the use, any other relevant facts reflecting differences in costs of providing service to the classes, or any combination of such factors; and
- Agreements between the municipal seller and outside city or wholesale customers that specifically provide for how rates are charged govern the parties' relationship.

12.0 SUMMARY

The primary purpose of allocating costs to cost-causative components is to provide a basis for further distribution of costs of service to customer classes based on their individual class service requirements reflecting these same cost-causative parameters. Cost-of-service allocations provide the necessary guidance to help utilities determine how much system users should pay for wastewater utility service.

Within the wastewater utility industry, two basic cost-allocation approaches—the design-basis method and the functional-basis cost-allocation method—are used to develop cost-of-service rates. Each has advantages and disadvantages that should be considered when utility management determines the allocation method best suited to meet its needs. Notably, utility management may elect to use hybrid methodologies to enable cost allocations to take full advantage of available data and knowledge of system design and operation. The existing body of utility case law provides representative rate-setting and implementation guidelines that should be recognized by utility management and its legal counsel when establishing or modifying utility rate policies to avoid rate-setting pitfalls that could unnecessarily lead to legal action.

13.0 SUGGESTED READING

National Association of Regulatory Utility Commissioners. (1996). *Uniform system of accounts for class a wastewater utilities.*

7

Distribution of Costs of Service to Customer Classes

1.0 INTRODUCTION

Costs incurred by a utility are directly related to the services it provides, as described previously in Chapter 6. Therefore, the process of identifying service characteristics of the utility's customers and distributing costs in proportion to their service demands are critical steps in the development of equitable rates and charges. If all customers had the same service characteristics, the cost of serving each customer would be equal. In this hypothetical case, a utility's costs could be distributed evenly among users, resulting in equal rates for all customers. In reality, however, this is not the case; different types of users have different service characteristics and, therefore, the costs associated with serving each customer vary. Because it is not practical or, in most cases, possible to determine the cost of serving each customer individually, it is an accepted industry practice to group customers with similar service characteristics into a limited number of classes to determine cost responsibility and, ultimately, rates applicable to each class.

Chapter 6 presented approaches and issues related to the allocation of wastewater system revenue requirements to cost-causative components. This chapter discusses the process of further distributing the system costs to customer classes based on their service demands. The assignment of cost responsibility to each customer class follows a three-step process, as shown in Figure 7.1. First, customer classes are established and units of service are quantified for each class for the period that rates will be effective. The service characteristics analyzed should parallel the cost-causative factors identified by the utility (discussed in Chapter 6). Second, unit costs are developed for each cost component or service characteristic. The third and final step is the distribution of costs to customer classes, based on individual class service units and component unit costs. Once costs have been identified for each customer class, rates may be developed to recover the assigned cost of service. Rate design is the subject of Chapter 8.

2.0 CUSTOMER CLASSES

2.1 Classification Factors

A careful analysis is necessary when establishing the appropriate classification of customers to ensure that resulting customer classes consist of users with similar use characteristics served at similar costs. Typically, three factors are used to determine the appropriate classification of a utility system's customers: (1) general service requirements, (2) customer use characteristics, and (3) geographic location.

Service requirements vary among customers, for instance, depending on whether they are receiving retail or wholesale service from the utility. A wholesale customer may, for example, own and maintain its own sewer collection mains and lift stations. Therefore, in this case, its service and cost responsibility would be limited to the utility's larger interceptor, pumping, and water resource recovery facilities (WRRFs). The utility's retail customers, on the other hand, may use the entire conveyance system. In this case, separate classes are necessary to equitably assign costs to customers based on their specific service requirements. Table 7.1 lists different types of service requirements provided by utilities with a description of each.

As discussed in Chapter 6, customer use characteristics, such as wastewater quantity and strength, are important cost-causative components of a wastewater system because systems must be sized to handle both peak

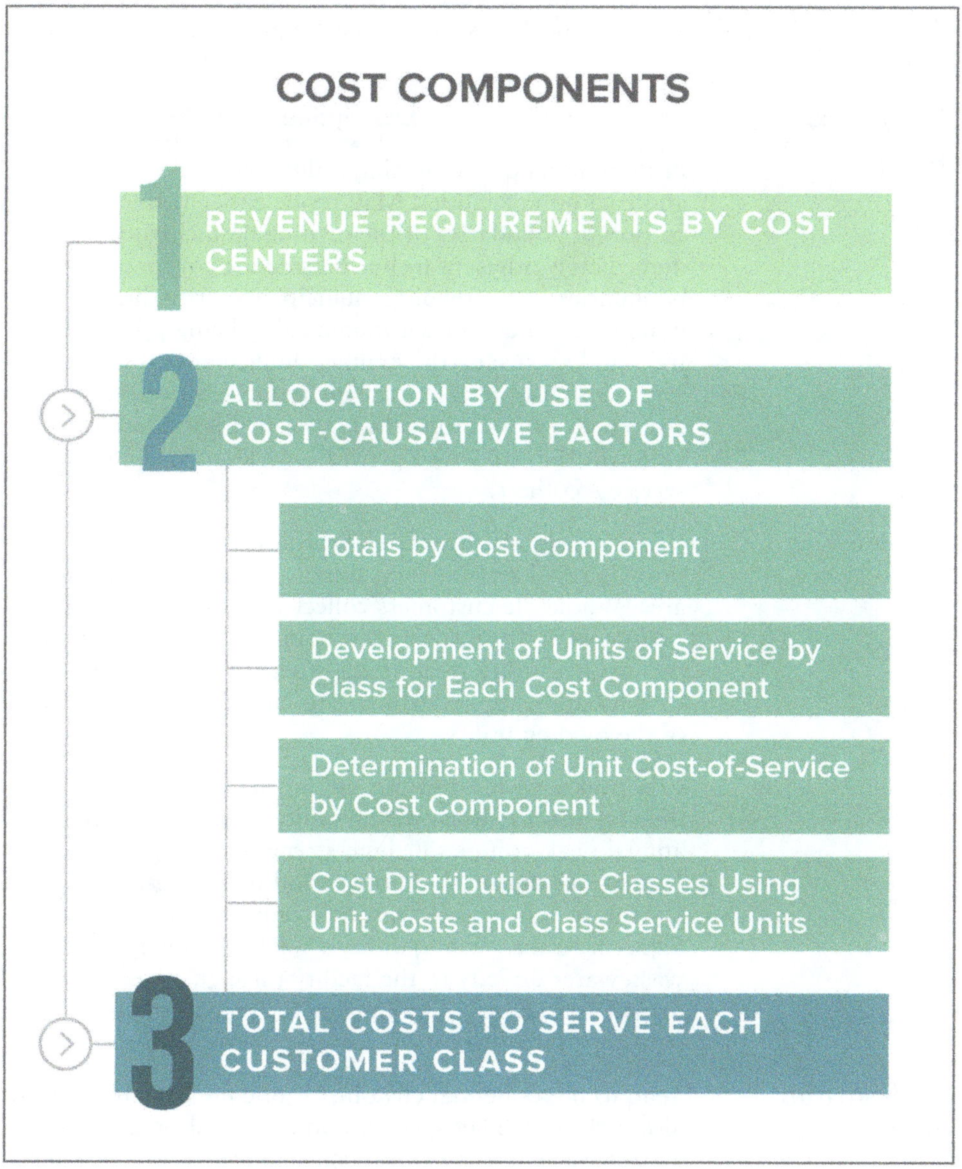

FIGURE 7.1 Cost-Allocation Flowchart

and average flows and treat wastewater of varying strengths and composition. Therefore, classification of wastewater customers should also consider volume, capacity, and strength characteristics of different types of customers' flows.

The third factor that may be considered when establishing customer classes is geographic location. This is predominantly used when service is provided to multiple cities or jurisdictions. One basis for differentiating

TABLE 7.1 Example of Different Types of Service Requirements Provided by Utilities

Service type	Description
Retail	Primary users of the owning utility's collection system and backbone facilities. May reside either within corporate boundaries of the owning municipality, immediately adjacent to but outside the corporate boundaries of the owning municipality, or within the boundaries of a contract municipality being provided full retail service by the owning municipality. Typically, investor-owned utilities provide retail service to their customers without regard to municipal boundaries, although they may also provide wholesale and contract service to others.
Wholesale	Use primarily the backbone facilities, especially if they provide retail service to customers in their own service area. Wholesale customers collect wastewater from their customers and transport it to points of connection with the owning utility treating the wastewater. Typically, these points of connection are to large interceptor mains of the owning utility.
Seasonal	Contribute wastewater during only a portion of a year, thereby affecting the owning utility with potentially annual peak volumes or high-strength loadings of wastewater during the short period that they use the system. In some cases, seasonal contributors are located adjacent to the owning utility's WRRF and may deliver wastewater directly to the facility through a customer-owned service main, minimizing the customer's use of backbone facilities.
High-strength haulers	Tend to be residential customer septic tank waste haulers, although some may specialize in industrial pretreatment wastes. Such customers' service is limited to wastewater treatment and disposal.
Pretreatment	Typically, industries that are required to treat their wastewater on site to remove pollutants to specified levels before they may discharge resulting lower-strength wastewater to the owning utility's system. Establishing an industrial pretreatment program is a requirement for municipal utilities that received federal construction grants. Typically, the owning utility monitors pretreatment customers' activities and tests their discharges to the utility system for a fee paid by such customers.

(continued)

TABLE 7.1 Example of Different Types of Service Requirements Provided by Utilities (*Continued*)

Service type	Description
Laboratory testing	Provided by the owning utility to periodically monitor strengths of wastewater entering its system by its wholesale, commercial, and industrial customers. Data are used by the utility to make treatment adjustments so that its discharges to U.S. waters meet the utility permit requirements. Some utilities provide laboratory testing services for some customers or other municipalities for a fee.
Combined sanitary and stormwater service	Typically, pertains to older municipal systems that have constructed combined disposal systems before changes in federal environmental law requiring either separation of the systems or treatment of all discharges. Typically, such stormwater flows would affect a wastewater utility more significantly with peak volumes than would solely infiltration/inflow in a utility system.
New customers or service areas	Often a wastewater utility has system capacity in excess of current peak requirements to meet the needs of future customers. As a result, either the wastewater utility must construct, with its own resources, or new customers and developers must construct or pay the utility through system development charges or front foot benefit fees to construct extension sewers to serve new customers or areas. If tax revenues and front-end capital cost recovery charges are not available, costs associated with extra capacity must be collected in the owning utility's service charges or through a special or betterment assessment for capital-cost recovery.

customers geographically is with respect to corporate limits of the city or jurisdiction. Such a distinction may be appropriate if the utility incurs different costs when serving users located outside of the corporate limits, compared to serving similar users inside the city or jurisdiction limits. Also, differences in the owner–customer relationship between inside and outside city users may justify a different rate or rate approach between these two groups. Other geographical considerations include elevation and distance. Elevation factors may be appropriate to consider in cases where pumped wastewater requires significant additional costs relative to nonpumped service, and pumped service is provided to discrete portions of the service area (e.g., pump zones). Distance (typically, in relation to the WRRF) may also be used because it relates to the length of pipe required and, therefore, the cost of service. For example, in situations where a municipal customer connects to an interceptor for transport of their wastewater to the WRRF, if the municipal customer only uses a portion of the interceptor, then that municipality may be allocated their proportionate share of costs on only the shared portion of the interceptor. This is known as the "reach method." However, when service is not provided to multiple cities or jurisdictions, the distance to WRRFs or which specific facility specific users contribute to are not considerations for cost-of-service differentiation among customers or customer classes. Table 7.2 provides a summary of typical classification factors.

2.2 Classification Approaches

The appropriate classification system for a utility must balance equity and fairness considerations with other concerns, including administrative

TABLE 7.2 Summary of Classification Factors

Factor	Sample types
General service requirements	Retail Wholesale Seasonal High-strength haulers Pretreatment, laboratory testing Combined sanitary and stormwater service New customer or service area
Usage characteristics	Wastewater flow • Customer contributed (volume and capacity) • Infiltration/inflow Wastewater strengths • Biochemical oxygen demand • Total suspended solids • Total Kjeldahl nitrogen • Phosphoros Customers Meter size
Geographic location	Inside or outside city Elevation (e.g., pump zones) Distance (e.g., from WRRF)

feasibility, revenue adequacy and stability, data availability, and public acceptance. Implementation of a highly disaggregated customer classification system may be complicated by the general problem of limited data to support customer class definitions and added administrative costs.

Historically, the typical approach to retail customer classification for wastewater utilities consisted of separate groups for residential, commercial, and industrial. For the purposes of the explanations and examples included in this chapter, the most typical retail customer classes are used, which are residential, commercial, and industrial. However, it should be noted that utilities may identify additional retail customer classes, such as multifamily residential or institutional, within their system and could include these as part of a cost-of-service analysis. Also, if applicable, classes for other types of services may be included, such as those listed in Table 7.1 (i.e., wholesale customers). Typically, residential customers include single-family dwellings, including mobile homes and trailers. Typically, commercial users consist of business enterprises such as motels, hotels, restaurants, and general office and retail establishments. Typically, industrial customers consist of manufacturing and processing establishments. Many exceptions to this "typical" practice are in place, often reflecting the prominence of particular types of users in a utility service area. For instance, some communities may establish a separate class for institutional customers, although these customers tend to be included in the commercial or industrial classes based on the nature of their activities. In addition, multifamily customers may be grouped as a separate class or combined with other classes (i.e., commercial or residential).

A simplified approach—consisting of relatively few customer classes—has been traditionally viewed as acceptable for the following reasons: (1) rates were relatively low and (2) limitations on data made it difficult to further disaggregate customer classifications by strength contributions of different types of users. Both of these conditions are changing, affording utility managers and policymakers the opportunity to advance rate equity, in some cases, through development of cost-of-service rates across a larger, better-defined set of customer classifications.

In communities where the diversity of the customer base warrants recognition of relative-strength contributions, typically, the following two approaches are used: (1) extra-strength surcharges and (2) strength-based classifications. A hybrid approach may also be appropriate. The extra-strength approach is common among utilities that are required by law to set up pretreatment programs for specified industrial users and by some state agencies to monitor potential high-strength users such as laundries, restaurants, and food processors. These users often have discharge characteristics that vary from typical domestic contributions, both in strength and the ratio between various constituents, which makes it unreasonable to allocate costs solely on a volume basis. In these cases, it is more appropriate for identified and monitored high-strength users to pay a surcharge for strengths above domestic-strength wastewater (U.S. Environmental Protection Agency, 1984). With this methodology, charges are developed on a per-constituent basis (i.e., flow, biochemical oxygen demand [BOD], total suspended solids [TSS], total Kjeldahl nitrogen [TKN], and phosphorus) based on the allocation of costs described in Chapter 6. The class-service requirement for this group is determined from the actual discharge, in volume or load depending on the constituent, and the strengths related to surcharge customers are removed from the total for determining the class-service requirement for all other classes.

Another approach is to assign commercial customers to separate classes, for example, to high, medium, and low wastewater strengths, or by particular types of businesses (i.e., restaurants, bakeries, or general office). Typically, classifications are based on the BOD and/or TSS concentrations for the various types of commercial customers. Similar to the surcharge method, costs are then allocated to customer classes from their flow and loading contribution using the per-constituent allocation discussed in Chapter 6. Most utilities have relatively little site-specific data on the strength contributions of industrial and commercial customers who are not subject to industrial waste monitoring. As a result, typically, assignment of customers to separate classes based on wastewater strengths will require expanded sampling efforts, supplemented with information from other communities or from standards developed by regulatory agencies or other sources. Similarly, many utilities do not have data on or the data collection procedures necessary to monitor building uses of nonresidential customer accounts. As a result, establishment of separate classes for commercial customers requires a comprehensive review, supplementing of current customer account records, and development of procedures to keep such information current. Table 7.3 provides examples of the extra-strength, strength-based, and hybrid classification systems. Typically, expansion of sampling programs and data collection procedures result in added administrative costs associated with a strength-based classification approach. In contrast, the extra-strength surcharge approach relies on readily available data from the industrial-monitoring program. The hybrid approach incorporates standardized strength data for certain high-strength-type customers (i.e., laundries and restaurants) to disaggregate commercial customers beyond a single class. From an equity perspective, the strength-based and hybrid approaches may be desirable when significant diversity exists beyond those customers subject to monitoring requirements. The cost effects need to be weighed against the community's equity and other objectives in determining the optimal classification approach.

TABLE 7.3 Sample Wastewater Customer Classification Approaches

Approach	Sample classes
Extra-strength surcharge	Residential
	Commercial
	Industrial
	Extra-strength
Strength-based classifications	Residential
	Commercial*
	Low-strength (combined BOD and TSS of 400 mg/L or less)
	Medium-strength (combined BOD and TSS of 400 to 800 mg/L)
	High-strength (combined BOD and TSS of 800 to 1200 mg/L)
	Very high-strength (combined BOD and TSS of more than 1200 mg/L)
Hybrid approach	Residential
	Commercial
	Industrial
	Standardized-strength class
	Extra-strength surcharge

*Strength thresholds by subclass are for illustration only.

3.0 UNITS OF SERVICE

Once customers have been classified into relatively homogeneous groups or classes, a common basis for identifying cost responsibility is established. Chapter 6 described the process of allocating wastewater system revenue requirements to cost components based on cost-causative factors, including wastewater flow (both contributed average and peak wastewater flow and inflow and infiltration [I/I]), wastewater strengths, customer service, and other cost components (e.g., collection system acreage and density). The next step in the rate development process is to quantify the service demands associated with each cost component for the period rates will be effective and identify each customer class's respective units of service. Costs by component may then be allocated to each class in proportion to service demands.

4.0 WASTEWATER VOLUMES

Wastewater volumes consist of two components: (1) customer-contributed wastewater and (2) I/I. Contributed wastewater volume is the quantity discharged directly from customers to the wastewater system. Depending on system characteristics and the cost-allocation methodology used, contributed wastewater volume may consist of average annual volume and peak volumes. Typically, because direct metering of wastewater volume is limited to large industrial or wholesale customers, it is common practice to use metered water data to estimate customer-contributed average wastewater volume units of service for the majority of customer classes.

A customer's metered water consumption can be used to estimate corresponding sewer use. However, in recognition of the often significant amount of water used by residential customers for lawn irrigation, car washing, or other purposes not discharged to the wastewater system, it is reasonable, typically, for allocation purposes, to estimate contributed residential wastewater volumes based on winter water use (typically, when irrigation is limited to summer months) or as a representative percentage of total annual use (when irrigation is year-round). Typically, contributed volumes for nonresidential customers are based on actual water sales. Sometimes, utilities may grant an allowance for product or process water that is not returned to the sewer where there is a defensible rationale for doing so (e.g., based on reliable documentation provided by the customer). For large users, the utility may also install a sewer meter to measure flows returned to the sewer. Some commercial accounts (e.g., schools) may have separate water meters for outdoor use so that the domestic water meter reflects water returned to the sewer.

Customer classes may deliver peak wastewater volumes to the system at different times. Wastewater volume from residential customers, for example, is at its peak on weekday mornings typically, as residents prepare for the workday. Residential loads may also peak in the winter in some parts of the country when faucets are turned on to keep pipes from freezing. Industrial customers tend to discharge peak volumes during the shutdown and cleanup of a process run. The timing and frequency of such events vary by industry. Commercial use tends to peak seasonally or because of a significant influx of convention visitors to the community. Individual class wastewater peaking estimates may be based on estimates of noncoincidental class maximum month or maximum day water demands (excluding allowances for irrigation and other nonsewered use). However, the lack of specific data on wastewater peak volumes by customer class tends to limit the practice of developing individualized class capacity factors.

The second element of flow that combines with contributed wastewater to make up the total volume reaching the WRRF is I/I. Infiltration is the flow entering the sanitary sewer resulting from high groundwater or precipitation that occurred days or weeks before the observed flow in the sanitary sewer. Inflow results from rainfall that enters the sanitary collection system through a number of direct connections such as catch basins, roof drains, foundation drains, and manhole covers. The total I/I in the system may be estimated based on available studies or comparisons of contributed wastewater and metered facility flows (see Table 7.4 and the facility balance discussion later in this section).

Many utilities operate separate sanitary sewer systems. Separate sanitary sewer systems have separate sanitary sewer and stormwater pipes. Sanitary sewer pipes carry wastewater to the WRRF for treatment, whereas the stormwater management system includes separate pipes, does not transport stormwater to the WRRF, and instead discharges collected stormwater to nearby surface waters. Separate sanitary sewer systems have varying degrees of I/I. Other utilities operate combined systems, where shared pipes transport both sanitary and stormwater flows to the WRRF. Utilities that operate combined systems may recognize a greater proportion of the total flows received at the facility as I/I compared to separate sanitary sewer systems.

Inflow and infiltration costs pose a special challenge in wastewater ratemaking because these costs are not a consequence of directly measurable service demands by utility customers. Groundwater levels, age of pipe, and soil conditions may influence the amount of I/I that enters the system from different basins. Administering a system of wastewater charges that differs by drainage basin, age of pipe, or soil conditions would be difficult and costly. Therefore, cost allocation approaches must be based on factors that estimate service requirements to equitably distribute I/I cost responsibilities.

As part of the 1972 Water Pollution Control Act (known as the Clean Water Act) and subsequent amendments, the U.S. Environmental Protection Agency issued guidelines regarding the establishment of wastewater rates. Included in these guidelines are sections related to recovery of operation and maintenance costs through user charges. Relevant portions of these guidelines discuss the allocation and recovery of I/I costs. The guidelines state that I/I costs shall be allocated and recovered using one or a combination of approaches that are identified. These approaches include recovery of I/I costs from customers in proportion to the following:

- Contributed wastewater volumes;
- Number of connections;
- Land area; and
- Property valuation of users, if the system has approval for user charges based on ad valorem taxes.

None of these methods fully captures the extent of I/I in the system, but each method has a relationship to the quantity of I/I. The most common approaches have been to use contributed wastewater flow, the number of

TABLE 7.4 Water Resource Recovery Facility Balance Analysis

Line no.	Category	(Volume 1000 Ccf)	(mil. gal)	Wastewater strength					
				TSS		BOD		TKN	
				(1000 lb)	(mg/L)	(1000 lb)	(mg/L)	(1000 lb)	(mg/L)
1	Total volume received at facility	15 740	11 774	28 567	291	22 727	231	3112	32
2	Less								
3	Total billed volume	11 084	8291						
4	Infiltration/inflow[a]	4656	3483	436	15	436	15	436	15
5	Surcharge strength	—	—	5225	4321	444			
6	Standardized strength[b]	1653	1236	6187	600	3609	350	412	40
7	Balance	—	—	—	—	—			
8	Domestic strength wastewater[c]	9431	7054	16 719	284	14 361	244	1820	31

[a]Assume I/I is the difference between facility inflow and billed volumes and has nominal strengths of 15 mg/L for TSS, BOD, and TKN.
[b]Assume standardized strength class represents 25% of commercial and industrial classes' volumes and customers.
[c]Assume average domestic-strength wastewater volumes come from residential, surcharge, and 75% of commercial and industrial customers.

connections (or customers), or a combination of the two to allocate I/I-related costs. Contributed flows are often used because of the correlation between flow volume and pipe size. Furthermore, volume recognizes the potential for greater runoff from large properties than smaller properties, and it is assumed that large-volume users have larger properties than small-volume users. Runoff from properties contributes to street runoff that has the potential to enter poorly sealed manhole covers, serving large-diameter mains or curb inlets directly to combined stormwater and sanitary sewers. Also, considerable inflow may enter large mains over time when they are placed in the groundwater table or below waterway crossings. The number of connections in the wastewater system reflects the size of the collection system; the greater potential for infiltration through poor joints and cracked pipes; and the potential for inflow through roof, floor, and foundation drains.

The property valuation approach is seldom used because few systems have an approved user-charge structure based on ad valorem taxes. A shortfall of the property area approach is that there is not always correlation between the area of the property and the length of lines that serve the property. In addition, the approach selected may be complicated by additional data requirements.

The costs of extra system capacity to convey, store, and treat wet weather peaks associated with I/I should be allocated as closely to their sources as possible. Separating the collection system into zones or sewer basins and analyzing wet weather flows on a per-basin basis is often the closest one can come to such an allocation. However, prorating I/I charges to certain zones of the system based on their contribution may be difficult because of additional data requirements or political issues surrounding higher charges to some communities and not others.

For systems where there are significant costs related to I/I or stormwater for combined sewer systems, the allocation approach selected could have a significant effect on the relative utility rates of different customer classes. Allocating I/I or stormwater costs in proportion to contributed wastewater volumes will result in more costs being recovered from large users, all other things equal, than if the number of connections is used to allocate costs.

5.0 WASTEWATER STRENGTHS

The wastewater strength component may be separated into costs that may vary with pollutants such as BOD, chemical oxygen demand, TSS, TKN, and phosphorus facility loadings. Water resource recovery facility influent records provide information on total pollutant loadings for the system. Typically, individual service units for each customer class are estimated from a combination of data, including sampling data of individually monitored customers (typically industrial and some wholesale and contract customers) and industry and other communities' data on average strength concentrations by type of customer. It is also proper to provide recognition of the pollutants present in I/I or stormwater that enter the sanitary system.

6.0 CUSTOMER SERVICE

Typically, customer-related costs include customer billing and service connection costs and, in some cases, meter reading and maintenance costs. Typically, billing costs are distributed to customer classes based on the number of bills. Whereas wastewater charges are based on quantity of water use, a portion of water-meter-related costs may be distributed to wastewater system customers. Typically, equivalent meter ratios are used to recover meter and service connection costs to reflect the increasing nature of such costs with meter and service-line size, whereas billing costs are recovered on a per-bill basis reflecting computer billing. However, if a large retail or wholesale customer requires special billing and collection considerations, a miscellaneous fee or direct assignment of cost may be used for special services that others do not receive.

7.0 UNUSED CAPACITY CHARGES

Depending on the methodology for the allocation of capacity to large users, utilities may not fully recover the costs associated with unused capacity (i.e., capacity that has been allocated to individual customers but is not consistently used). Although the costs to maintain this capacity are less than when the capacity is used, the utility does bear a carrying cost to provide this capacity. If this cost is not considered in the distribution of customer costs, the utility may be unable to fully recover revenue.

This scenario is often present in utilities with a large industrial customer base (as a percentage of total flow or loads) that does not require pretreatment to near municipal levels. Often, capacity is allocated to these large industrial customers above and beyond their typical use to allow for increased production or seasonal variations. In these cases, it may be appropriate to impose a charge for unused capacity based on the utility's cost of maintaining this capacity. It is not appropriate or reasonable to apply this methodology to small customers. However, this cost-recovery methodology has proven effective for large industrial customers as a means of recovering revenue (Gregg et al., 2014).

8.0 OTHER UNITS OF SERVICE

In some systems, special recognition should be given to industrial waste sampling or the recovery of costs associated with required pretreatment program monitoring. Special cost components and customer units may be required to properly establish customer class responsibility.

In addition, units of service related to acreage or density are sometimes used for allocation of collection system costs. When customers within classes are relatively close together, the investment in collection system costs is less than when those customers who are relatively spread out. Thus, density, or the ability to serve more customers per length of pipe has a significant effect on the level of collection system operating and capital costs.

9.0 FACILITY BALANCE ANALYSIS

As discussed previously, typically, the projections of customer class units of service are based on limited local data, supplemented with data from other communities. As a means of "testing" the assumptions and data used in the analysis, particularly as they relate to estimation of customer class volume and units of service, a facility balancing technique may be used. This technique involves comparing the total units of service estimated from customer classes to the actual flows and loads measured at the facility. Through this analysis, assumptions of strength concentrations, water-use adjustments, and I/I may be refined to improve the overall equity of the cost-allocation process and ensure that rates are based on a reasonable representation of actual volumes to the facility, discouraging potential claims of overcharging insofar as billed units of service could exceed actual units of service.

Table 7.4 provides an example of a facility balance analysis. In the example provided, total facility volume (line 1) reduced by total billed wastewater volume (line 3) to estimate I/I (line 4). The I/I is assumed to have a nominal strength loading of 15 mg/L for TSS, BOD, and TKN. Total facility TSS, BOD, and TKN loadings are reduced further by monitored "surcharge" industrial loadings (line 5) and assumed standardized commercial and industrial strengths (line 6). Ideally, wastewater strength assumptions would be supported by local sampling data. However, when local data are not available, industry reference sources may be used, for example, the California State Water Resources Control Board's *Revenue Program Guidelines* (Appendix G from the Policy for Implementing the State Revolving Fund for Construction of Wastewater Treatment Facilities [1998].) The balance (line 7) is attributed to the remaining customers that include residential, other commercial and industrial customers not included in the standardized class, and the portion of surcharge customer load that is average strength.

10.0 SUMMARY OF CUSTOMER CLASS UNITS OF SERVICE

Table 7.5 shows units of service by customer class under the design-basis cost-allocation methodology described in Chapter 6. Service units under the design basis include volume, capacity (peak wastewater volume), strength (suspended solids, BOD, and TKN), and customer billing units. The volume units by class (line 1 in Table 7.5) come from customer billing records, and I/I volume is estimated based on the facility-balance analysis (Table 7.4).

As shown in Table 7.5, capacity units of service (line 5) are estimated based on wastewater volumes and peak-to-average-day capacity factors. Capacity factors for contributed wastewater for this illustration are estimated to be 1.5 for all classes. The capacity factor for I/I is 4.0.

Wastewater strength loadings (lines 9, 10, and 11) are based on class volumes and the following: assumed strengths for I/I and standardized strength customers, monitored strengths for surcharge customers, and estimated domestic strengths from the facility-balance analysis for remaining customers.

The I/I loadings (estimated in Table 7.4) are redistributed to the customer classes, two-thirds by the number of customers and one-third by the amount of class-contributed volume. For example, I/I volume is estimated to be 4 656 000 Ccf (1 Ccf = 100 cu ft = 748 gal), as indicated in Table 7.4 (line 4). Two-thirds (3 104 000 Ccf) is redistributed back to the customer classes in proportion to each class's share of total customers, and the other one-third (1 552 000 Ccf) is redistributed in proportion to each class's wastewater volume. Residential customers represent approximately 86.4% of total customers (net of surcharge customers) and 40.4% of wastewater volumes. The residential class is, therefore, responsible for 3 306 000 Ccf of I/I volume, as shown in the following process and in lines 2 and 3 of Table 7.5:

In U.S. customary units:

Total I/I allocated 2/3 by customers × Residential class percentage of total = Residential class I/I by customers

3 104 000 Ccf × 0.864 = 2 680 000 Ccf rounded

Total I/I allocated 1/3 by volume, 1000 Ccf × Residential class percentage of total = Residential class I/I by volume, 1000 Ccf

TABLE 7.5 An Example of Customer Classes' Units of Service in Design-Basis Cost-Allocation Methodology

Line no.	Description	Residential	Commercial	Industrial	Surcharge	Standardized strength	Total
	Wastewater volume (1000 Ccf[a])						
1	Billed wastewater volume	4473	2761	2197		1653	11 084
2	I/I allocated 2/3 by customers	2680	282	36		106	3104
3	I/I allocated 1/3 by volume	626	387	308		231	1552
4	Total annual treated volume	7779	3430	2541	0	1990	15 740
	Wastewater capacity flowrate (Ccf/d[b])						
5	Estimated wastewater peaks	18 382	11 347	9029		6793	45 551
6	I/I allocated 2/3 by customers[c]	29 378	3087	391		1159	34 015
7	I/I allocated 1/3 by volume[c]	6864	4237	3371		2537	17 009
8	Total	54 624	18 671	12 791	0	10 489	96 575
	Wastewater strength (1000 lb)						
9	TSS	7930	4895	3895	5225	6187	28 131
10	BOD	6811	4204	3346	4321	3609	22 291
11	TKN	863	533	424	444	412	2676
12	I/I allocated 2/3 by customers[d]	753	79	10	0	30	872
13	I/I allocated 1/3 by volume[d]	176	109	86	0	65	436
	Customer billing units						
14	Number of bills	208 400	21 900	8328	828	24 672	264 128
15	Number of customers	52 100	5475	694	69	2056	60 394

[a]1 Ccf = 2.83 m^3.
[b]1 Ccf/d = XX unit of measure.
[c]Infiltration/inflow volumes represent one-third or two-thirds of the volume from Table 7.4 and a 4.0 peaking factor.
[d]Infiltration/inflow strengths represent one-third or two-thirds of the combined total pounds for TSS, BOD, and TKN, shown in Table 7.4.

$$1552\ 000 \text{ Ccf} \times 0.4035 = 626\ 000 \text{ Ccf, rounded}$$

Total residential class I/I by customers, 1000 Ccf + Total residential class I/I by volume, 1000 Ccf

$$= 2\ 680\ 000 \text{ Ccf} + 626\ 000 \text{ Ccf} = 3\ 306\ 000 \text{ Ccf}$$

The preceding process is used to allocate I/I loadings for each service parameter and customer class. The results are shown in Table 7.5 (lines 2 to 3, 6 to 7, and 12 to 13). As alluded to in footnote b of Table 7.5, in the interest of saving space for a nominally small portion of the overall system units of service, the combined strength of I/I for TSS, BOD, and TKN (each assumed to have a strength of 15 mg/L) are shown on lines 12 and 13 of Table 7.5.

Table 7.6 shows the units of service by customer class under the functional-basis cost-allocation methodology. Under this approach and for purposes of this example, capacity is not used as a cost determinant. Volume, strength, and customer units of service are determined in the same manner as described previously for the design-basis cost-allocation methodology.

11.0 UNIT COSTS AND DISTRIBUTION OF COSTS TO CUSTOMER CLASSES

The significant effort in distributing costs to customer classes is complete after customers are classified and their respective units of service are determined. Determination of customer class revenue responsibility can now be established based on component unit costs and projected units of service for each customer class. Unit costs are calculated for each component based on allocated revenue requirements and total number of applicable units of service for the test period. The unit costs by component are then applied to individual class units of service to determine the cost responsibility of each class. This process is illustrated in Tables 7.7 and 7.8, with a summary of the two approaches provided in Table 7.9.

Table 7.7 illustrates calculation of unit costs of service and distribution of costs to customer classes based on the design-basis cost-allocation methodology described in Chapter 6. Under the design basis, costs are allocated to volume, capacity (peak wastewater volume), strength (TSS, BOD, and TKN), customer billing, customer, and industrial monitoring categories. Based on the allocated costs by component and the total units of service (from Tables 7.4 and 7.5), unit costs of service are calculated for each component. These unit costs are then used to distribute costs to customer classes based on the

TABLE 7.6 Customer Classes' Units of Service in Functional-Basis Cost-Allocation Methodology

Line no.	Description	Residential	Commercial	Industrial	Surcharge	Standardized strength	Total
1	Billed wastewater volume (1000 Ccf[a])	4473	2761	2197		1653	11 084
2	I/I allocated 2/3 by customers[b]	2680	282	36		106	3104
3	I/I allocated 1/3 by volume[b]	626	387	308		231	1552
4	Total annual treated volume	7779	3430	2541	0	1990	15 740
	Wastewater strength (1000 lb[c])						
5	TSS	7930	4895	3895	5225	6187	28 131
6	BOD	6811	4204	3346	4321	3609	22 291
7	TKN	863	533	424	444	412	2676
8	I/I allocated 2/3 by customers[d]	753	79	10	0	30 872	
9	I/I allocated 1/3 by volume[d]	176	109	86	0	65 436	
	Customer billing units						
10	Number of bills	208 400	21 900	8328	828	24 672	264 128
11	Number of customers	52 100	5475	694	69	2056	60 394

[a]1 Ccf = 2.83 m^3.

[b]Infiltration/inflow volumes represent one-third or two-thirds of the volume from Table 7.4.

[c]1 lb/? = XX mg/L.

[d]Infiltration/inflow strengths represent one-third or two-thirds of the combined total pounds for TSS, BOD, and TKN shown in Table 7.4.

TABLE 7.7 Unit Costs of Service and Allocation to Classes in Design-Basis Cost-Allocation Methodology

Line no.	Description	Total	Volume	Capacity	TSS	BOD	TKN	Customer billing	Customer	Industrial surcharge
	Annual costs									
1	Net operation and maintenance	$18,251,000.00	$1,223,000.00	$7,072,000.00	$2,338,000.00	$3,243,000.00	$1,321,000.00	$1,299,000.00	$1,099,000.00	$656,000.00
2	Total capital costs	$10,585,000.00	$1,082,000.00	$3,937,000.00	$1,842,000.00	$2,616,000.00	$792,000.00	$73,000.00	$243,000.00	$-
3	Totals	$28,836,000.00	$2,305,000.00	$11,009,000.00	$4,180,000.00	$5,859,000.00	$2,113,000.00	$1,372,000.00	$1,342,000.00	$656,000.00
	Annual units of service									
4	Quantity		$15,740,000.00	$96,575.00	$28,567,210.00	$22,727,210.00	$3,112,210.00	$264,128.00	$60,394.00	$828.00
5	Unit of measure		Ccf*	Ccf/d	lb	lb	lb	bills	each	bills
6	Unit costs of service (per unit)		$0.15	$113.99	$0.15	$0.26	$0.68	$5.19	$22.22	$792.27
	Classes of service									
	Residential									
7	Units of service		$7,779,000.00	$54,624.00	$8,239,000.00	$7,121,000.00	$1,173,000.00	$208,400.00	$52,100.00	
8	Allocated annual costs	$13,444,000.00	$1,140,000.00	$6,227,000.00	$1,205,000.00	$1,835,000.00	$797,000.00	$1,083,000.00	$1,157,000.00	
	Commercial									
9	Units of service		$3,430,000.00	$18,671.00	$4,957,000.00	$4,267,000.00	$595,000.00	$21,900.00	$5,475.00	
10	Allocated annual costs	$5,095,000.00	$502,000.00	$2,128,000.00	$725,000.00	$1,100,000.00	$404,000.00	$114,000.00	$122,000.00	
	Industrial									
11	Units of service		$2,541,000.00	$12,791.00	$3,927,000.00	$3,378,000.00	$456,000.00	$8,328.00	$694.00	
12	Allocated annual costs	$3,644,000.00	$372,000.00	$1,458,000.00	$575,000.00	$871,000.00	$310,000.00	$43,000.00	$15,000.00	
	Surcharge									
13	Units of service				$5,225,000.00	$4,321,000.00	$444,000.00	$828.00	$69.00	$828.00
14	Allocated annual costs	$2,842,000.00			$765,000.00	$1,114,000.00	$301,000.00	$4,000.00	$2,000.00	$656,000.00
	Standardized strength									
15	Units of service		$1,990,000.00	$10,489.00	$6,219,000.00	$3,641,000.00	$444,000.00	$24,672.00	$2,056.00	
16	Allocated annual costs	$3,811,000.00	$291,000.00	$1,196,000.00	$910,000.00	$939,000.00	$301,000.00	$128,000.00	$46,000.00	
17	Totals	$28,836,000.00	$2,305,000.00	$11,009,000.00	$4,180,000.00	$5,859,000.00	$2,113,000.00	$1,372,000.00	$1,342,000.00	$656,000.00

*1 Ccf = 2.83 m^3.

customer class units of service shown in Table 7.5. Surcharge customers are assumed to be allocated costs of the industrial surcharge program entirely.

Table 7.8 illustrates the development of units of service and distribution of costs to customer classes for the functional-basis cost-allocation methodology, also described in Chapter 6. As in Table 7.7, the process relies on customer billing units developed in Table 7.6 and unit costs of service derived from the allocated costs by component divided by total billing units. Also, industrial surcharge program costs are allocated entirely to surcharge customers.

Table 7.9 summarizes costs by customer class under the alternative allocation methodologies. As Table 7.9 indicates, residential and commercial customers are responsible for a greater share of costs under the design basis relative to the functional basis. This is because, under design-basis cost allocation, more costs are allocated to volume (both average volume and capacity) relative to strength components. Industrial-, surcharge-, and standardized-strength customers have greater cost responsibility under the functional-basis cost-allocation methodology because a larger portion of costs are allocated based on wastewater strengths. For this reason, the functional-basis cost allocation may be more appropriate in cases with large industrial or other users whose contributions to the system vary from typical residential contributions.

12.0 COMMUNAL AND SPECIFIC COSTS

In establishing unit costs and customer class costs of service, it may be necessary to consider differences in general service requirements among customers, as discussed previously in this chapter and in Chapter 6. Costs for a component may be spread across the total units of service in the case of "communal" costs, or only among a subset of customer service units in the case of "specific" costs. The allocation of industrial surcharge program costs directly to benefitting customers in Tables 7.7 and 7.8 is an example of a specific-cost allocation.

13.0 RETURN ON INVESTMENT

When utility service is provided to non-owner customers, such as wholesale or retail service customers located outside the jurisdiction of the owner municipality, it may be appropriate to recognize return on facility value related to providing such service. In this instance, the utility basis for determining cost of service, as discussed in Chapter 5 and more fully in Appendix A, is

TABLE 7.8 Unit Costs of Service and Allocation to Classes in Functional-Basis Cost-Allocation Methodology

Line No.	Description	Total	Volume	TSS	BOD	TKN	Customer billing	Customer	Industrial surcharge
	Annual costs								
1	Net operation and maintenance	$18,251,000.00	$5,042,000.00	$3,359,000.00	$4,189,000.00	$982,000.00	$1,313,000.00	$2,742,000.00	$624,000.00
2	Total capital costs	$10,585,000.00	$3,473,000.00	$2,468,000.00	$3,618,000.00	$710,000.00	$73,000.00	$243,000.00	$-
3	Totals	$28,836,000.00	$8,515,000.00	$5,827,000.00	$7,807,000.00	$1,692,000.00	$1,386,000.00	$2,985,000.00	$624,000.00
	Annual units of service								
4	Quantity		$15,740,000.00	$28,567,000.00	$22,727,000.00	$3,112,000.00	$264,128.00	$60,394.00	$828.00
5	Unit of measure		Ccf[a]	lb[b]	lb	lb	bills	customers	bills
6	Unit costs of service (per unit)		$0.54	$0.20	$0.34	$0.54	$5.25	$49.43	$753.62
	Classes of service								
	Residential								
7	Units of service		$7,779,000.00	$8,239,000.00	$7,121,000.00	$1,173,000.00	$208,400.00	$52,100.00	
8	Allocated annual costs	$12,641,000.00	$4,208,000.00	$1,680,000.00	$2,446,000.00	$638,000.00	$1,094,000.00	$2,575,000.00	
	Commercial								
9	Units of service		$3,430,000.00	$4,957,000.00	$4,267,000.00	$595,000.00	$21,900.00	$5,475.00	
10	Allocated annual costs	$5,042,000.00	$1,855,000.00	$1,011,000.00	$1,466,000.00	$324,000.00	$115,000.00	$271,000.00	
	Industrial								
11	Units of service		$2,541,000.00	$3,927,000.00	$3,377,000.00	$456,000.00	$8,328.00	$694.00	
12	Allocated annual costs	$3,662,000.00	$1,375,000.00	$801,000.00	$1,160,000.00	$248,000.00	$44,000.00	$34,000.00	
	Surcharge								
13	Units of service			$5,225,000.00	$4,321,000.00	$444,000.00	$828.00	$69.00	$828.00
14	Allocated annual costs	$3,422,000.00		$1,066,000.00	$1,484,000.00	$241,000.00	$4,000.00	$3,000.00	$624,000.00
	Standardized strength								
15	Units of service		$1,990,000.00	$6,219,000.00	$3,641,000.00	$444,000.00	$24,672.00	$2,056.00	
16	Allocated annual costs	$4,069,000.00	$1,077,000.00	$1,269,000.00	$1,251,000.00	$241,000.00	$129,000.00	$102,000.00	
17	Totals	$28,836,000.00	$8,515,000.00	$5,827,000.00	$7,807,000.00	$1,692,000.00	$1,386,000.00	$2,985,000.00	$624,000.00

[a] 1 Ccf = 2.83 m^3.
[b] 1 lb/? = ? mg/L.

TABLE 7.9 Comparison of Design-Basis and Functional-Basis Cost Allocation Methodology Results

Line no.	Customer classes	Design basis	Functional basis	Difference
1	Residential	$13,444,000.00	$12,641,000.00	$803,000.00
2	Commercial	$5,095,000.00	$5,042,000.00	$53,000.00
3	Industrial	$3,644,000.00	$3,662,000.00	$(18,000.00)
4	Surcharge	$2,842,000.00	$3,422,000.00	$(580,000.00)
5	Standardized strength	$3,811,000.00	$4,069,000.00	$(258,000.00)
6	Total	$28,836,000.00	$28,836,000.00	$—

particularly applicable because it recognizes the costs and risks utility owners must assume. As owners, the municipality or utility is entitled to a fair return on the value of facilities devoted to serving non-owner customers. Consideration of the cost of borrowed funds, business risk, level of service provided, and method used for determining facility value should be incorporated to the evaluation of an appropriate rate of return that provides adequate inducement to the utility to continue providing service to its non-owner customers.

14.0 SUMMARY

An equitable allocation of revenue requirements to wastewater system users begins with an analysis of customer service demands. Because it is not practical to charge each customer individually, customers must be classified into relatively homogeneous groups, such as residential, commercial, and industrial, with similar use characteristics. Costs are then allocated to customer classes in proportion to each class's service demands. For the wastewater system, costs are allocated to customers based on their general service requirements, wastewater discharge characteristics, number of customers, and other factors.

As wastewater rates continue to increase and as utilities strive to meet the challenges of the 21st century, greater recognition of differences in customer service demands through the implementation of more disaggregated customer classifications may be desirable in many communities. However, such classifications may be complicated by the general problem of limited data to support customer class definitions and the ongoing administrative requirements related to tracking changing customer uses. Therefore, the appropriate classification of customers must balance equity considerations with other concerns, including administrative feasibility, revenue adequacy and stability, data availability, and public acceptance.

Cost-of-service studies are important for determining the necessity for and level of rate adjustments to be considered in recovering costs equitably from wastewater system customers. The importance of such studies is that they make policymakers aware of all factors contributing to the costs incurred in providing service to various system customers. When practical or when policy matters dictate deviation from cost of service, decision-makers can be apprised of the financial effects of deviation.

15.0 REFERENCES

California State Water Resources Control Board. (1998, March). Revenue program guidelines (appendix G from the *Policy for implementing the state revolving fund for construction of wastewater treatment facilities*).

Gregg, M., Fuss, M., Mieyr, M., & Jarocki, W. (2014, September 27–October 1). A case study in innovative financial approaches: Utilizing an industrial incentive policy to achieve Nampa's strategic goals. *Proceedings of the 87th Annual Water Environment Federation Technical Exhibition and Conference* [CD-ROM], New Orleans, LA, United States. Water Environment Federation.

U.S. Environmental Protection Agency. (1984). *User charge guidance manual for publicly-owned treatment works* (EPA 430/9-84-006).

16.0 SUGGESTED READINGS

Koplow, D., & Lownie, A. (1998). *Cost accounting and budgeting for improved wastewater treatment* (Final report). Industrial Economics, Incorporated.

Rothstein, E., & Galardi, D. (2000, October 14–18). Analysis of strength loading characteristics for wastewater classification. *Proceedings of the 73rd Annual Water Environment Federation Technical Exposition and Conference* [CD-ROM], Anaheim, CA, United States. Water Environment Federation.

8

Development and Design of a Schedule of Rates and Charges

1.0 INTRODUCTION

The "art" of rate-making involves designing rates that balance potentially conflicting objectives in a manner that reflects community values. At a minimum, utility rates should be sufficient to generate revenues required to support operations, develop capital infrastructure, and preserve (or enhance) the financial integrity of the utility system. Revenue generation should also provide for effective asset management and adequate renewal and replacement of aging infrastructure, which has historically been underfunded (graded C- in the American Society of Civil Engineers' *Report Card for America's Infrastructure* [2021]). Beyond technical considerations, rate design is frequently used to further specific policy objectives. For example, rate structures have been designed to support economic development and assistance to low- or fixed-income customers, among other objectives. Development of utility rates must also consider administrative requirements.

Chapter 7 discussed the distribution of costs to customer classes under a cost-of-service framework. This chapter describes the process of developing a schedule of rates and charges to recover allocated cost responsibilities from each customer class. This process consists of selection of the rate structure(s), use of customer billing units developed in Chapter 7, and calculation of rate and charge levels. Typically, charges referred to in this discussion are fixed amounts paid each billing period, regardless of whether service is rendered; rates are prices per quantities of service that apply to the variable components of service rendered during each billing period, such as estimated customer effluent collected, conveyed, and treated. Implementation issues are also discussed.

2.0 SELECTION OF RATE FORMS

Multiple rate structures may be implemented to generate a specific level of revenue from a particular customer class, the selection of which is based on the balancing of multiple rate objectives. Furthermore, each of the basic rate structures may be implemented on a systemwide basis or for individual customer classes. Mixing rate structures across customer classes is common and offers an opportunity to tailor rate design to address policy objectives related to specific customer classes. Rate equity may be enhanced through more complex rate structures that involve multiple components and rate classes. However, these potential equity gains must be evaluated against practical data and billing system limitations and implications for public understanding and acceptance.

The number and relative importance of rate policy objectives will vary from community to community, often reflecting socioeconomic conditions. As utilities engage in selecting a rate form, it is useful to explicitly identify their rate policy objectives and ultimately be prepared to review how the selected alternative is consistent with broader community policies. Sample rate-policy objectives are outlined briefly in Table 8.1.

Notably, cost-of-service rates are the standard for determination of rate equity. Comparisons to cost-of-service-based allocations of customer class cost responsibility provide information on the magnitude and direction of

TABLE 8.1 Sample Rate-Policy Objectives

Policy objective	Description
Revenue stability	The rate form limits annual and monthly variability in revenues to facilitate effective financial planning and limit rate adjustments precipitated by variances from revenue projections.
Administrative feasibility	The rate form may be accommodated by the utility's customer billing system and requires an acceptable level of administrative review and intervention to enable adequate revenue reporting and customer service.
Equity	The rate form provides for recovery of allocated costs of service by customer class.
Public understanding and acceptance	The rate form is well understood or understandable by the general rate-paying population and consistent with their perceptions of acceptability.
Affordability	The rate form provides wastewater service to meet basic health and human needs at a cost that is affordable for fixed- or low-income ratepayers.
Economic development	The rate form provides wastewater service at costs for commercial and industrial ratepayers that will help promote (or retain) business interests.
Water conservation	The rate form acts to (or reinforces water rate structures that) promote the efficient use of water resources (typically via higher unit costs for higher levels of water use).

subsidies across customer classes. This information is critical for evaluating the merit of deviations from cost-based rates and structuring charges on individual rate components to serve a community's policy objectives.

2.1 Rate Components

Typically, wastewater rate structures feature some combination of fixed and variable components, although, in some cases, a single charge or rate may be applied. (Fixed components are those that do not vary with the quantities of wastewater volumes and strengths contributed by a customer to the wastewater system. Variable components are directly dependent on flow volume or strength contributed.) Additional charges may be assessed to recover specific costs from customers within the service area or a subset of customers based on differences in service requirements. For example, typically, separate charges are developed for customers subject to industrial pretreatment or industrial waste monitoring and sampling regulation. In general, these charges are assessed on a unit-cost basis whereby customers are charged per pound of extra-strength of pollutant loading. Typically, wastewater strength is measured using biochemical oxygen demand (BOD) and total suspended solids (TSS). In some cases, chemical oxygen demand may be used in place of BOD. Other measurements that may be used include total Kjeldahl nitrogen (TKN) and phosphorus. As discussed in Chapters 6 and 7, rates for industrial surcharge customers may also reflect the direct allocation of industrial program costs to these customers. In addition, although less common, charges may be assessed to customers in a particular geographic area or basin where significant cost differences can be identified.

2.1.1 Fixed Charges

Fixed charges remain the same regardless of the volume or strength of wastewater discharged by a customer. The absence of wastewater service metering and the need to establish a stable revenue stream historically precipitated the use of fixed charges as the predominant rate form. Rate designs based on a single fixed charge component remain common today among small utility systems and in service areas without water metering. However, the use of fixed charges exclusively is steadily declining and being replaced by rate schedules that combine fixed charges with a volume rate. When used in conjunction with volume rates, fixed charges commonly take the form of (1) service charges, (2) meter charges, (3) minimum charges, or (4) combinations thereof.

The terms *service charge* and *customer charge* are often used interchangeably and refer to a charge that is applied equally to all customers in a given billing period. Although more common in water rate-making than wastewater rate design, a meter charge may be implemented that establishes a fixed fee that increases with the size of the customer's water meter (e.g., to recognize costs for increasing potential customer flow from different-sized connections).

Finally, a meter charge and service charge may be combined with a quantity allowance to establish a minimum charge. In this case, the fixed charge could recover all or a portion of volume-related costs (including inflow and infiltration [I/I] costs allocated on a customer basis) as part of the minimum and the customer and meter costs. The main advantage of a minimum charge approach is revenue stability. As more costs are recovered through a fixed charge, less revenue is subject to fluctuations because of customer use patterns. Primary disadvantages of revenue recovery largely through fixed charges are potential inequities in the distribution of cost responsibilities and lack of affordability for small-volume users (i.e., customers whose monthly use is below the quantity allowance). Payment for quantities or services not received is contrary to cost-of-service principles. To the extent that small users may have less flow than the quantity allowance, they would be charged in excess of their cost of service. Although the amount of this excess may be a small dollar amount, it may be significant to low- and fixed-income households.

2.1.2 Volume Rates

Volumetric rates, which may be referred to as *commodity*, *flow*, or *variable rates*, are applied against a customer's estimated service volume (above any quantity allowances provided through minimum charges). Figure 8.1 offers examples of the most common volumetric rate structures. Typically, a uniform volume rate is the most used rate form for wastewater service, although declining block rates have also been used. A uniform volume rate form assesses a single rate per unit of service volume for all flow volumes. In contrast, a declining block rate form applies successively lower rates to successively increasing levels of flow, or volumetric blocks.

Declining block rates have been viewed somewhat more favorably in the past because of the economies of scale characteristic of utility infrastructure construction. Increasing the capacity of utility facilities did not always impose proportionate increases in costs, so lower rates for higher volumetric flow were viewed as consistent with utility cost structures. However, over at least the last decade, there has been a rather significant movement away from declining block rates in favor of uniform volume designs (and, in some cases, inclining block designs for utilities that desire consistency with their water rate structure). The trend toward uniform volumetric rates is often considered to be consistent with a current emphasis on equity in cost recovery from a broad base of use patterns, expressing the idea that all contributed wastewater volumes should equivalently share in cost recovery.

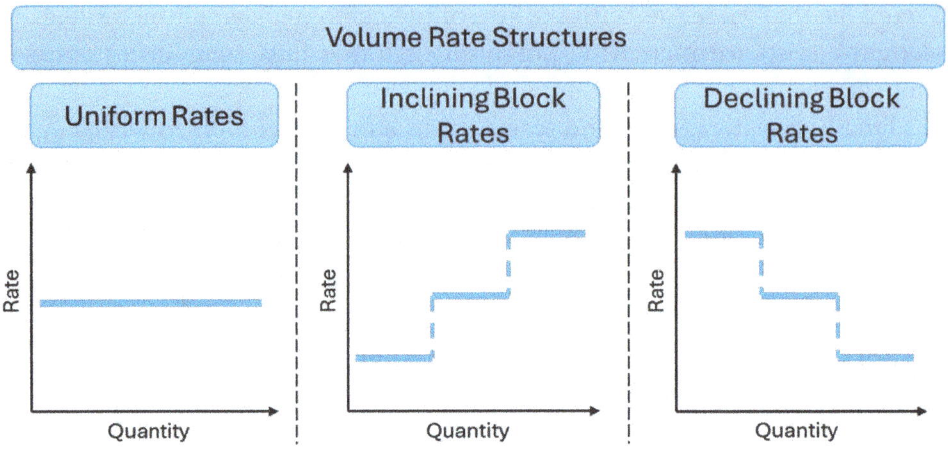

FIGURE 8.1 Examples of the Most Common Volumetric Rate Structures

2.1.3 Constituent Rates

Whereas volume rates are almost always used to recover costs of average-strength wastewater, individual constituent rates are used for recovering extra-strength waste loads from monitored commercial and industrial customers. To the extent that components of a customer's use of a utility system can be identified and costs can be allocated for handling constituent contributions, rates by constituent may be established. For example, industrial waste customers' contributions of BOD and TSS above specified concentrations are often surcharged. In this case, industrial waste customers' flows are regularly sampled, and BOD and TSS surcharges are calculated by applying a charge-per-weight of excess contribution.

2.1.3.1 Advantages

The advantages of this form of constituent rate are that equity may be enhanced in charges applicable to a group of industrial and commercial customers whose diversity defies development of precise customer class definitions. Wide variances in costs to serve based on significant differences in constituent loadings are addressed directly.

2.1.3.2 Disadvantages

Constituent rates require the collection of customer-specific data on strength contributions. In the case of industrial waste surcharges, wastewater sampling is already required to obtain sufficient data for process control and monitoring. If not required for regulatory compliance, it may be difficult to justify the significant costs for this data collection. Additionally, some consider constituent rates as adding to the complexity of customer billing and administration.

2.2 Rate Evaluation Matrix

A rate evaluation matrix may be used to summarize the performance of the existing rate structure and alternative structures being considered regarding the utility's policy objectives. In some cases, the evaluation matrix may be used to provide a qualitative, narrative review of policy considerations, as illustrated in Table 8.2. In other cases, utilities have used a scoring system to rank alternative rate forms and facilitate communication of their selections.

3.0 DETERMINATION OF BILLING UNITS

Billing units are an essential component to developing wastewater cost of service, calculating rates and charges, and assessing appropriate user charges to customers. Yet, determining suitable wastewater billing units for customers is not straightforward and, consequently, can be accomplished by several different approaches.

3.1 Customer Fixed Units

There are two most typically accepted approaches to determining the number of customer units, or fixed units. These approaches are outlined in the following subsections.

3.1.1 Units by Meter/Account

The most common approach to determining wastewater fixed charges is based on a per-account basis. Typically, an account is related to one water meter. This is the easiest determinant for identifying a wastewater customer and is frequently used as the sole basis for the application of fixed charges.

3.1.2 Units by Meter Size

Some utilities assess base charges that scale up by the size of the customer's water meter, as discussed previously in this chapter. The customer is still assessed only one fixed charge, but it is dependent on the potential load from different-sized customers as reflected by the size of the water meter. In this case, the utility calculates equivalent units, or customers, based on water meter capacity ratios reflecting the differences in potential demand for each meter size. This is common practice for water service because considering system and customer peaking is an important factor in sizing the system.

TABLE 8.2 Rate-Evaluation Matrix

Policy objective	Fixed charges	Uniform volume rate (all classes)	Uniform volume rate (all classes)	Inclining block volume rates by class with customer charges	Inclining block volume rates by customer class	Notes
Administrative feasibility	Simple. May or may not be class-differentiated, but no billing of volumes.	Simplest, not requiring customer class designation.	Simple, though customer class designation required.	Customer charges may or may not include a quantity allowance.	Requires billing capability and evaluation of appropriate block thresholds.	Inclining blocks and quantity allowances in customer charges will require bill frequency distribution data.
Revenue stability	Most stable; affected only by changes in number of customers.	Directly tied to stability of billing determinants and overall price elasticity of demand.	Some insulation from revenue instability via differential rates by class.	Inclusion of fixed customer charges enhances revenue stability.	Relatively less stable due to revenue recovery in higher flow blocks.	In general, dependent on stability of billing determinants. If all metered water use, likely less stable than winter average.
Equity	No recognition of volume-based differences in costs to serve within customer classes.	No recognition of cost-of-service differentials by customer class.	Simplest of cost-of-service rate forms; no recognition of fixed versus variable aspects of costs.	Most direct cost-of-service rate form, provides recognition of cost causation of available bill determinants.	To extent wastewater costs increase with customer-induced flow capacity, may further enhance equity. Given wet weather flow issues, this correlation is tenuous.	Equity, for purposes herein, refers to consistence with cost-of-service allocations of cost responsibility.

Public understanding and acceptance	Easily understood. Acceptance dependent on recognition of cost-of-service inequities.	Most easily understood, though inequities may result in challenges.	Easily understood while also enabling implementation of cost-of-service rates for each class.	Relative complexity vs implementation of cost-of-service rates and closer matching of charges to cost causation.	Limitations on customer discretion in wastewater use may result in challenges.	Highly dependent on level of community participation and interest in wastewater rate-making and policy issues. As rates rise, this objective will be of increasing importance.
Affordability	Disadvantages low-volume, low- and fixed-income users, and those who might reduce water use to make bills affordable. May benefit higher volume users.	Absence of customer charge may be advantageous, though lower-strength classes subsidize high-strength users.	Absence of customer charge may be advantageous, though no specific benefit for low- or fixed-income users.	Customer charges may be accompanied by subsidized quantity allowance to address affordability.	Subsidized initial block for low-volume residential use may address some affordability concerns. However, low- or fixed-income but high volume users (because of household size and leaking plumbing) are adversely affected.	Rate policies may only support redress of low- or fixed-income affordability challenges. Typically, there is a correlation between low- or fixed-income and low volume; however, subsidized rates (as a blunt instrument) also benefit wealthy low-volume users. To address affordability concerns, implementing targeted complementary conservation or assistance programs may be merited.

However, for wastewater service, there is less of a direct nexus between the meter capacity and wastewater system use because of other factors, such as average day volumes, peaking resulting from I/I, or wastewater loadings that are more likely to drive investment. Because of this, determining billing units by customers' water meter sizes is less common than uniform per-account charges. An alternative is the development of a measure of equivalent use to reflect average daily volumes by different meter sizes or for different classes of customers. Such equivalent residential unit based systems are sometimes used to recover fixed costs for different sizes of customers and can provide an equitable basis for cost recovery.

3.2 Customer Volumetric Units

Determining the customer's volumetric units, or wastewater flow, is more complicated than establishing the customer's account. In most cases, customers are not metered for wastewater flow. In some cases, large-volume or wholesale customers may have metered wastewater flow, but, even then, wastewater meters tend to present reliability issues for the utility. Therefore, in most cases, customer wastewater flow is estimated using one of several approaches.

3.2.1 Metered Water Use

The most straightforward determination of a customer's wastewater units is by directly linking to the customer's water consumption, which is determined by the customer's monthly water meter reads. For example, if a residential customer uses 10 Ccf (7480 gal) of water in a month, then the customer will be billed for wastewater flow of 10 Ccf. Although this information is most readily available, an underlying assumption is that all metered water is discharged to the wastewater system for treatment, which is most often not the case because of outdoor watering and other practices.

3.2.2 Winter Averaging

Another approach that is commonly implemented to correct the shortfall of directly linking to metered water consumption is winter averaging. Under this approach, a customer's wastewater billable units are determined by calculating the mean, or average, of the customer's metered water consumption during a designated series of months, typically during the winter months (e.g., December, January, and February) when outdoor watering is at a minimum. The winter average is an attempt to quantify the amount of water discharged to the wastewater system and treated, which is less than the monthly metered consumption typically. The customer is then billed their unique winter average for each month during the subsequent year until their new winter average is determined during the following winter period.

3.2.3 Percentage of Metered Water Consumption

Similar to winter averaging, another approach that is not as common is billing as a percentage of a customer's metered water consumption. For example, a residential customer may use 10 Ccf (7480 gal) of water in 1 month, but the wastewater utility may choose to bill only 90% of the customer's metered use, or, in this example, 9 Ccf of wastewater flow. Again, this is a mechanism to address the concern that not all metered water consumption is discharged to the wastewater system.

4.0 DEVELOPMENT OF THE RATE SCHEDULE

Once the rate form(s) have been selected and the billing units are estimated, the rate schedule can be developed. The rate schedule is composed of a set of rates and charges, typically segregated by customer class and, under certain circumstances, rates and charges that will be assessed to individual customers. Rates and charges applicable to customer classes are calculated by dividing the revenue to be recovered from each rate component by class by the number of associated billing units by class estimated for the rate-setting period (typically a year). Similarly, rates and charges applicable to individual customers are calculated by determining unit costs for specific services.

4.1 Systemwide Uniform Volume Rates

Using the information developed in Chapter 7 and presented in Tables 7.5 and 7.7, many different rate forms can be calculated relatively easily. For example, the simplest of rate forms requires only the rate revenue requirement, less revenues recovered through industrial waste surcharges, and billable flows (11 084 000 Ccf, from Table 7.5; 1 Ccf = 2.38 m^3). For the design-basis cost-allocation methodology, in which surcharge revenues are $2,842,000, a rate for all customer classes of $2.35/Ccf is calculated by dividing the net revenue requirements ($28,836,000 − $2,842,000 = $25,994,000) by billable flows. Using the functional-basis cost-allocation methodology, the calculation results in a systemwide uniform rate of $2.29/Ccf because the surcharge revenues are $580,000 more than under the design-basis cost-allocation methodology.

In the event that a single uniform volume rate, with a customer charge, is desired, the calculation is slightly more complex. Using the design-basis cost-allocation methodology shown in Table 7.7, allocated billing costs of $1,368,000 (note that an additional $4,000 of customer billing costs are recovered through industrial waste surcharges) are divided by the number of bills issued across all customer classes (263,300, from Table 7.5, excluding surcharge customer bills) to determine a customer billing charge of $5.20 included in all bills. Allocated customer costs of $1,340,000 (note that an additional $2,000 of customer billing costs are recovered through industrial waste surcharges) are divided by the number of customers (accounts) across all customer classes (60,325 from Table 7.5, excluding surcharge customers) divided by 12 months to determine a customer charge of $1.85 per month. These should be combined to determine the total fixed charge of $10.75 for quarterly billed customers and $7.05 for monthly billed customers. A separate customer charge would be calculated for surcharge customers that also included their allocated costs of the surcharge program ($656,000 from Table 7.7). The volumetric rate is similarly calculated as all allocated volume- and strength-related costs (excluding industrial waste surcharge costs) of $23,286,000 divided by billed flows of 11 084 000 Ccf for a rate of $2.10/Ccf. Using the functional-basis cost-allocation methodology (Table 7.8) and the same calculation procedure and billing units as in the design-basis example, the total fixed charge is $17.61 for quarterly billed customers and $9.37 for monthly billed customers; the volumetric rate is $1.90 per Ccf. These calculations are presented in Table 8.3.

4.2 Industrial Waste Surcharges and Pretreatment

Note that the rate revenue requirement used above is less than the total revenue requirement of $28,836,000. This difference reflects the assumption that industrial waste surcharges will be assessed on a unit-cost basis, as indicated in Table 7.7. That is, using the design-basis cost-allocation methodology, industrial waste surcharge customers will be charged $801/bill ([$656,000 allocated industrial surcharge costs + $6,000 allocated customer billing costs to surcharge customer] divided by 828 surcharge billings), plus the following:

- $0.146/lb (1 lb = 0.4536 = kg) for TSS,
- $0.258/lb (1 lb = 0.4536 = kg) for BOD, and
- $0.678/lb (1 lb = 0.4536 = kg) for TKN,

where the amounts of pollutant loading are determined by sampling of individual industrial waste customers' discharges.

TABLE 8.3 Calculation of Systemwide Uniform Volume Rates Using Both Design- and Functional-Basis Cost-Allocation Methodologies

	Uniform volume rate—all classes (design basis)	Uniform volume rate—all classes (functional basis)	Uniform volume rate with customer charge (design basis)	Uniform volume rate with customer charge (functional basis)
Allocated costs				
Customer billing			$1,368,000	$1,382,000
Customer			$1,340,000	$2,982,000
Flows and strengths			23 286 000	21 050 000
Total	$25,994,000	$25,414,000	$25,994,000	$25,414,000
Billing units—number of bills			263 300	263 300
Billing units—number of accounts			60 325	60 325
Billing units—volume	11 084 000 Ccf	11 084 000 Ccf	11 084 000 Ccf	11 084 000 Ccf
Billing charge—per bill			$5.20/bill	$5.25/bill
Customer charge—per account per month			$1.85/account	$4.12/account
Fixed charge—quarterly billed customers			$10.75/account	$17.61/account
Fixed charge—monthly billed customers			$7.05/account	$9.37/account
Volume rate	$2.35/Ccf	$2.29/Ccf	$2.10/Ccf	$1.90/Ccf

Note. 1 Ccf = 2.83 m^3.

4.3 Uniform Rates by Customer Class

For the development of cost-of-service-based rates, differentiation of rates by customer class is required. Uniform volume rates by customer class provide a mechanism to do so, achieving cost recovery either through a combination of a customer charge and volumetric rate or simply by a single rate-per-unit of billable flow. Again, using the information developed in Chapter 7, uniform volume rates may be readily calculated as shown in Table 8.4 for the design-basis cost-allocation methodology.

Note that, although cost allocations to service characteristics consider total flows, billed flows must be used for rate calculations to achieve full cost recovery. For the combined rate structure, volume- and strength-related costs allocated to each customer class are divided by billed flows of each class to determine the volumetric rate. Customer and billing costs are recovered through the fixed quarterly or monthly charge. The uniform volume rate without customer charge rate form reflects the addition of all costs allocated to a customer class divided by billed flows. For the functional-basis cost-allocation methodology, uniform volume rates by customer class are presented in Table 8.5.

For those rate designs that include a fixed charge, additional revenue stability may be achieved by recovering some of those costs allocated to volume and strength parameters through the customer charge. For example, I/I costs allocated in proportion to customers are often recovered through the fixed-charge component.

The "standardized-strength" class provides an illustration of the disaggregation of the traditional customer class structure retained by many utilities. As shown in Table 7.4, this class of customers (that is not subject to industrial waste surcharges) is assigned strength characteristics that are significantly above those associated with domestic-strength wastewater. The service rates for this class, as shown in Tables 8.4 and 8.5, reflect the additional cost to serve customers contributing higher-strength wastewater.

Implementation of a standardized-strength class may be considered for purposes of enhancing the equity of wastewater rates. Doing so allows reflecting the higher-strength wastewater contributions of many commercial customers not subject to industrial waste surcharges. As indicated in Chapter 7, some communities have implemented multiple-strength classes (e.g., high strength, medium strength, low strength) rather than a single "standardized-strength" class to recognize the different strength-loading characteristics of their diverse commercial classes. Other communities have isolated particular user types (e.g., restaurants) to recognize differences in costs to serve.

TABLE 8.4 Uniform Volume Rates by Customer Class Rate Design Using Design-Basis Cost-Allocation Methodology

		By class with customer charge			By class without customer charge		
	Classes of service	Cost recovery ($)	Billed units	Unit charge	Cost recovery ($)	Units	Unit charge
	Fixed charges						
1	Billing charge ($/bill)	1,368,000	263 300	$5.20			
2	Customer charge ($/month)	1,340,000	60 325	$1.85			
3	Fixed charge ($/quarterly bill)			$10.75			
4	Fixed charge ($/monthly bill)			$7.05			
	Residential						
5	Fixed charge ($/quarterly bill)			$10.75			
6	Volume rate ($/Ccf*)	11,204,000	4 473 000	$2.50	13,444,000	4 473 000	$3.01
	Commercial						
7	Fixed charge ($/quarterly bill)			$10.75			
8	Volume rate ($/Ccf)	4,859,000	2 761 000	$1.76	5,095,000	2 761 000	$1.85
	Industrial						
9	Fixed charge ($/monthly bill)			$7.05			
10	Volume rate ($/Ccf)	3,586,000	2 197 000	$1.63	3,644,000	2 197 000	$1.66
	Standardized strength						
11	Fixed charge ($/monthly bill)			$7.05			
12	Volume rate ($/Ccf)	3,637,000	1 653 000	$2.20	3,811,000	1 653 000	$2.31
13	Totals ($)	25,994,000			25,994,000		
14	Surcharge	2,842,000			2,842,000		
15	Check	28,836,000			28,836,000		

*1 Ccf = 2.83 m^3.

TABLE 8.5 Uniform Volume Rates by Customer Class Rate Design Using Functional-Basis Cost-Allocation Methodology

	Classes of service	By class with customer charge			By class without customer charge		
		Cost recovery ($)	Billed units	Unit charge	Cost recovery ($)	Units	Unit charge
	Fixed charges						
1	Billing charge ($/bill)	1,382,000	263 300	$5.25			
2	Customer charge ($/month)	2,982,000	60 325	$4.12			
3	Fixed charge ($/quarterly bill)			$17.61			
4	Fixed charge ($/monthly bill)			$9.37			
	Residential						
3	Fixed charge ($/quarterly bill)			$17.61			
4	Volume rate ($/Ccf*)	8,972,000	4 473 000	$2.01	12,641,000	4 473 000	$2.83
	Commercial						
7	Fixed charge ($/monthly bill)			$17.61			
8	Volume rate ($/Ccf)	4,656,000	2 761 000	$1.69	5,042,000	2 761 000	$1.83
	Industrial						
11	Fixed charge ($/monthly bill)			$9.37			
12	Volume rate ($/Ccf)	3,584,000	2 197 000	$1.63	3,662,000	2 197 000	$1.67
	Standardized strength						
15	Fixed charge ($/monthly bill)			$9.37			
16	Volume rate ($/Ccf)	3,838,000	1 653 000	$2.32	4,069,000	1 653 000	$2.46
17	Totals ($)	25,414,000			25,414,000		
18	Surcharge	3,422,000			3,422,000		
19	Check	28,836,000			28,836,000		

*1 Ccl = 2.83 m^3.

Notably, while disaggregation of customer classes may afford improvements in the equity of cost distributions, there are important implementation considerations and limitations. First, as with any customer classification process, some intraclass differences in specific customer strength characteristics will remain. For example, some customers in a particular class may contribute higher BOD loads and lower TSS loads, or vice versa, to those assigned the class as a whole. (Although, if these variances are widespread, consideration may be given to structuring of the customer classes to better match classes with strength-loading characteristics of customer bases.) While equity improvements may be realized by recognition of strength loading beyond industrial waste surcharges, classifications will, by no means, be precise.

Second, the classification process requires that information on customer building uses be collected, analyzed, and maintained. Although typical strength-loading characteristics of different types of building uses are readily available, class disaggregation will require the utility to identify building uses of its commercial customer base, assign strength-load parameters, and restructure its customer classes. In addition, provisions will need to be established to maintain this information as building uses change over time.

Third, typically, it is appropriate to afford customers the opportunity to request reclassification by demonstrating that strength-load assignments are invalid for their particular account. Accordingly, appeal procedures are required that define how customers may secure reclassification.

Finally, the increased complexity of the customer classification system will impose public communication and administrative requirements. Commercial customers, in particular, should be apprised of the basis for classifications and their opportunities for appeal. Similarly, customer service representatives will be required to both communicate this information and ensure the appropriate tracking of customer building uses in the utility's billing system.

4.4 Bill Impact Analysis

For any of the rate forms selected, it is important to understand the potential effects on ratepayer bills that will result from a rate structure change. Accordingly, a basic bill impact analysis is indispensable, particularly for customer service representatives and public information officers. This form of analysis provides a calculation of bills under existing and proposed rates. Importantly, it reveals the differential intraclass effects of rate changes when cost recovery changes between fixed and volumetric charges. To illustrate this for a single customer class, Table 8.6 provides an example of a bill impact analysis for residential customers, who are assumed to

TABLE 8.6 Bill Impact Analysis for Residential Customers Using Design-Basis Cost-Allocation Methodology and Proposed Rates From Table 8.4

				Current rate			Proposed rate	
	Customer charge (quarterly)			$7.00			$10.75	
	Volume rate ($/Ccf*)			$2.70			$2.50	
	Current bills			Proposed bills			Total monthly wastewater bill impact	
Quarterly billed flows	Quarterly fixed charge	Quarterly volume charge	Total quarterly bill	Quarterly fixed charge	Quarterly volume charge	Total quarterly bill	Amount	Percentage
1	$7.00	$2.70	$9.70	$10.75	$2.50	$13.25	$3.55	36.65%
2	$7.00	$5.40	$12.40	$10.75	$5.01	$15.76	$3.36	27.09%
3	$7.00	$8.10	$15.10	$10.75	$7.51	$18.26	$3.16	20.96%
4	$7.00	$10.80	$17.80	$10.75	$10.02	$20.77	$2.97	16.68%
5	$7.00	$13.50	$20.50	$10.75	$12.52	$23.27	$2.77	13.53%
6	$7.00	$16.20	$23.20	$10.75	$15.03	$25.78	$2.58	11.12%
7	$7.00	$18.90	$25.90	$10.75	$17.53	$28.28	$2.38	9.20%
8	$7.00	$21.60	$28.60	$10.75	$20.04	$30.79	$2.19	7.65%
9	$7.00	$24.30	$31.30	$10.75	$22.54	$33.29	$1.99	6.37%
10	$7.00	$27.00	$34.00	$10.75	$25.05	$35.80	$1.80	5.29%
11	$7.00	$29.70	$36.70	$10.75	$27.55	$38.30	$1.60	4.37%
12	$7.00	$32.40	$39.40	$10.75	$30.06	$40.81	$1.41	3.57%
13	$7.00	$35.10	$42.10	$10.75	$32.56	$43.31	$1.21	2.88%
14	$7.00	$37.80	$44.80	$10.75	$35.07	$45.82	$1.02	2.27%
15	$7.00	$40.50	$47.50	$10.75	$37.57	$48.32	$0.82	1.73%
16	$7.00	$43.20	$50.20	$10.75	$40.08	$50.83	$0.63	1.25%
17	$7.00	$45.90	$52.90	$10.75	$42.58	$53.33	$0.43	0.82%

18	$7.00	$48.60	$55.60	$10.75	$45.09	$55.84	$0.24	0.43%
19	$7.00	$51.30	$58.30	$10.75	$47.59	$58.34	$0.04	0.07%
20	$7.00	$54.00	$61.00	$10.75	$50.10	$60.85	($0.15)	−0.25%
21	$7.00	$56.70	$63.70	$10.75	$52.60	$63.35	($0.35)	−0.55%
22	$7.00	$59.40	$66.40	$10.75	$55.11	$65.86	($0.54)	−0.82%
23	$7.00	$62.10	$69.10	$10.75	$57.61	$68.36	($0.74)	−1.07%
24	$7.00	$64.80	$71.80	$10.75	$60.12	$70.87	($0.93)	−1.30%
25	$7.00	$67.50	$74.50	$10.75	$62.62	$73.37	($1.13)	−1.52%
26	$7.00	$70.20	$77.20	$10.75	$65.12	$75.87	($1.33)	−1.72%
27	$7.00	$72.90	$79.90	$10.75	$67.63	$78.38	($1.52)	−1.90%
28	$7.00	$75.60	$82.60	$10.75	$70.13	$80.88	($1.72)	−2.08%
29	$7.00	$78.30	$85.30	$10.75	$72.64	$83.39	($1.91)	−2.24%
30	$7.00	$81.00	$88.00	$10.75	$75.14	$85.89	($2.11)	−2.39%
31	$7.00	$83.70	$90.70	$10.75	$77.65	$88.40	($2.30)	−2.54%
32	$7.00	$86.40	$93.40	$10.75	$80.15	$90.90	($2.50)	−2.67%
33	$7.00	$89.10	$96.10	$10.75	$82.66	$93.41	($2.69)	−2.80%
34	$7.00	$91.80	$98.80	$10.75	$85.16	$95.91	($2.89)	−2.92%
35	$7.00	$94.50	$101.50	$10.75	$87.67	$98.42	($3.08)	−3.04%
36	$7.00	$97.20	$104.20	$10.75	$90.17	$100.92	($3.28)	−3.14%
37	$7.00	$99.90	$106.90	$10.75	$92.68	$103.43	($3.47)	−3.25%
38	$7.00	$102.60	$109.60	$10.75	$95.18	$105.93	($3.67)	−3.35%
39	$7.00	$105.30	$112.30	$10.75	$97.69	$108.44	($3.86)	−3.44%
40	$7.00	$108.00	$115.00	$10.75	$100.19	$110.94	($4.06)	−3.53%
		Total Quarterly Bill (Current)				Total Quarterly Bill (Proposed)		

*Ccf = 2.83 m³.

have been previously charged non-cost-of-service rates of $7.00 per month plus $2.70/Ccf. The proposed rates are the uniform volume rate and fixed charge for all customer classes determined using the design-basis cost-allocation methodology provided in Table 7.7 and discussed previously. For this example, the shift in revenue recovery from volumetric rates to customer effects will adversely affect the lowest volume users, although the absolute dollar value of bill impacts is limited. However, in many communities, a shift in revenue recovery from fixed charges to variable rates has been made to support water conservation objectives or simply enhance ratepayers' control of wastewater bills. Figure 8.2 provides a snapshot of customer bills at 4 usage levels to more easily show the relationship of fixed versus variable for different use levels.

For communities that elect to implement cost-of-service-based rates in the context of increasing revenue requirements, it is often useful to calculate what portion of the bill impacts are required to achieve full revenue recovery versus the bill impact associated with movement to cost-based rates.

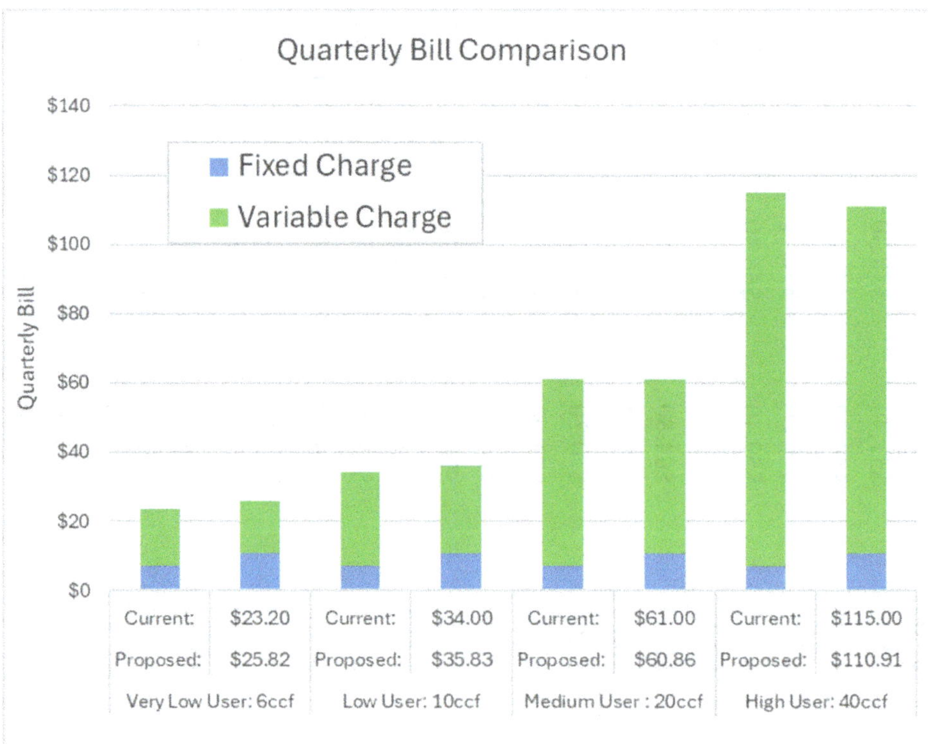

FIGURE 8.2 Residential Customer Quarterly Bill Comparison of Fixed Versus Variable Charges

5.0 WHOLESALE AND CONTRACT CUSTOMERS

Contracts may be desirable when wastewater service is furnished to unusually large-volume customers, customers discharging excessively high- or non-standard-strength wastewater, or customers who discharge on a seasonal or fluctuating basis. The wastewater utility must have the hydraulic and treatment capacity to handle these loads when they occur, but if the excess loadings or volumetric demands fluctuate, this capacity then remains idle the rest of the time. Unless the utility designs and implements a cost-of-service rate for these customers, it will either operate at a loss or be forced to recover these costs in the rates charged to its other customers, thus subsidizing the large or otherwise nonstandard customer. To develop a cost-of-service-based rate, operating and capital costs associated with serving such customers should be developed at the same time and by the same cost-of-service allocation process used to determine cost recovery for all other classes of service. A rate between the utility and the customer can be designed that will recover the necessary revenues on a year-round basis and the resulting rate can be incorporated to the utility's schedule of rates and charges specifically for this customer. Non-rate-cost recovery mechanisms can be appropriate in some cases, for example, flat monthly charges, budget allocation processes, and cost allocations with true-ups.

Wholesale service to other communities or districts is also frequently provided by wastewater utilities and the associated rate-design process must take into consideration the level of service provided to the customer. As discussed in Chapter 7, if the wholesale customer owns and operates its collection system and provides its own billing services, it should only be billed for transporting wastewater from the connection point to the water resource recovery facility and for handling and treating the wastewater.

6.0 POLICY-BASED ADJUSTMENTS

Rates based on cost of service provide a benchmark for evaluating the effects of more policy-driven approaches. To the extent that these policy approaches result in interclass or intraclass subsidies, it is important to be able to quantify these transfers and gauge customers' potential responses. A public involvement program that provides opportunities for community members to actively evaluate and provide input on approaches will generate rate structures that integrate policy-based adjustments with technical considerations.

6.1 Low-Income Affordability

As utility rates continue to increase nationwide, concerns about the affordability of service, particularly for low-income customers, are prevalent in many communities. One option for addressing this concern has been the implementation of "lifeline" rates, whereby income-qualifying ratepayers are eligible for a lower rate to ensure affordability of basic services. Fixed charges that include a quantity allowance that is priced below actual cost of service have also been used to address affordability concerns, although such allowances produce significant complications because the reduced costs for minimum service are provided to all ratepayers.

The question of affordability is particularly challenging to address in the context of establishing systemwide or customer class rates. Typically, low-income customers are not readily identifiable from customer billing records so, typically, are not segregated into separate customer classes. In the absence of class identification, rates that provide discounts to low-income users will necessarily result in somewhat higher charges to other users to fully recover all costs. In some communities, the provision of affordable service has been attempted through underpricing of low-volume use. Although this may help ensure the affordability of minimum use, this practice may not achieve the desired goals because a correlation between low-income customers and low-volume use is by no means assured. A further discussion of the critical issue of low-income affordability, including both rate design and programmatic approaches, is provided in Chapter 12.

6.2 Economic Development and Business Retention

Discounted wastewater rates have been used as one of many inducements for siting decisions for industrial and commercial facilities or for helping to discourage facility relocations. This rate policy deviates from cost-of-service principles and contemplates subsidization of selected business interests by other utility ratepayers in exchange for other tangible benefits. Explicit economic development rates are relatively uncommon and, typically, are placed in effect for a limited term. If used, these rates often require negotiation with affected parties to ensure that community economic development objectives will be advanced in exchange for utility revenue losses.

6.3 Rate Transitioning

In many instances, the implementation of cost-based rates will result in significant shifts in cost responsibilities within and across customer classes. In this case, utilities must weigh the equity gains of an immediate shift to cost-based rates, with issues related to affordability and customer budgeting challenges. Typically, a multiyear rate-transitioning period can accomplish both objectives by phasing in cost-based rates over a limited time to allow customers time to plan and adjust to the new rates and avoid "rate shock." However, such a policy will result in rates being below the cost of service for some users and above the cost of service for others during the transition period, which will perpetuate inequities embedded in current rates.

6.4 Water Conservation Efforts

Water conservation effects on traditional "domestic"-strength effluent (e.g., 250 mg/L) are affecting treatment at water resource recovery facilities and increasing costs for wastewater utilities. Many residential and commercial properties have or are starting to install low-flow/low-flush, water-saving fixtures, the effects of which are increasing traditional discharge-strength effluent and raising treatment costs on a per-unit-volume basis that may not be appropriately allocated among customer classes. As more communities enhance their conservation efforts, wastewater utilities may want to monitor the effects and consider reevaluating cost of service appropriately.

7.0 IMPLEMENTATION ISSUES

As a practical matter, selection of any form of utility rate design requires consideration of a variety of implementation issues in addition to technical and policy issues.

7.1 Data and Billing System Constraints

Perhaps the single most important practical rate design constraint is the absence of broadly available, accurate, customer-specific data on service use. Whereas long-distance telephone utilities collect data on each call made, including location, duration, and period (e.g., peak vs. off peak), wastewater utilities are limited in the type and frequency of data collected. Typically, direct measurement of wastewater flows and strengths are limited to relatively few large customers, and indirect data from water metering is limited to monthly or bimonthly intervals. As a consequence, wastewater rates must be established wherein the billing determinants are restricted to volume of use (as estimated from water meter data) and customer account information (e.g., type of customer and meter size) and imputed strength-loading characteristics.

7.2 Ordinances

Typically, utility policies related to wastewater use, rates, and billing for wastewater service are contained in ordinances, including the limits of service and rates charged to outside city customers or suburban units. Provisions for enforcing discharge requirements regarding high-strength wastes, toxins, and heavy metals are also included. Such requirements should be reviewed for compliance with current regulations set by local, state, and federal agencies.

8.0 SUMMARY

Ultimately, utility rates provide price signals about the cost of utility services and customers' relative responsibilities for the use of these services. To the extent possible, utilities should always strive to implement cost-of-service-based rates to minimize any interclass subsidies. In certain situations, however, policy-driven approaches are applied and likely result in interclass or intraclass subsidies. Rates based on cost of service can provide a benchmark for evaluating the effects of these policy-driven strategies, and it is important for a utility to understand, both qualitatively and quantitatively, the effects of policy-driven approaches.

As a practical matter, selection of any form of rate design requires consideration of a variety of implementation issues. For example, theoretically sound rate designs are effectively irrelevant if data are unavailable to support determination of customer bills.

Rate designs should also reflect community values because rate designs that are appropriate for one community may be inappropriate for another community.

9.0 REFERENCE

American Society of Civil Engineers. (2021). *Report card for America's infrastructure*. www.infrastructurereportcard.org

9

Wet Weather Financing and Cost Recovery

1.0 INTRODUCTION

In the context of wastewater systems, the term *wet weather flows* refers to rainfall- and snowmelt-induced extraneous "inflows" that are collected and conveyed by a combined sewer system (CSS), by a municipal separate storm sewer system (MS4), or as direct surface discharge to receiving waters. The characteristics of wet weather flows and their potential environmental, economic, regulatory, and community effects are different from that of groundwater inflow and infiltration (I/I) that occur naturally in a wastewater system. From a wastewater system perspective, the collection and conveyance of wet weather flows in a wastewater system is not a new phenomenon. In the discussion of cost-causative factors in Chapter 6, wet weather flows are deemed a part of I/I, and guidelines are provided to allocate I/I costs to volumetric- and customer-cost components.

However, many wastewater utilities, especially those with a CSS, grapple with significant unfunded regulatory mandates in the form of long-term control plan (LTCP) requirements and consent decrees to manage wastewater system wet weather flows. Both CSS's and MS4s involve significant financial investment in the treatment and management of wet weather flows (Environmental Finance Advisory Board, 2020). In utilities that have an integrated network of CSS and separate storm sewer system infrastructure, such as in Newark, New Jersey; Philadelphia, Pennsylvania; Toledo, Ohio; and Seattle, Washington, the magnitude of operation and maintenance (O&M) and capital costs of wet weather flow management and the recovery of those costs add an additional layer of complexity in wastewater rate-making.

In such cases, the more common approach of recovering all wastewater utility costs through sewer rates and charges poses various challenges, including funding adequacy, nexus between cost recovery and demands on the system, and equity in cost recovery. Therefore, with escalating regulatory and infrastructure management costs, many utilities recognize the need for alternative forms of funding and cost recovery that not only reflect principles of equitable recovery, fairness, and revenue stability, but also provide fee reduction opportunities for private and public on-site wet weather flow management for longer-term resilience.

In this chapter, wet weather financing and cost recovery are discussed primarily in the context of integrated wastewater utilities that have responsibilities for both sewer and stormwater management services. The following are the key questions that are pertinent to wet weather financing and cost recovery:

- What is the effect of wet weather flows on wastewater systems?
- What are some feasible funding and cost recovery mechanisms?

- What are the key considerations in wet weather cost recovery?
- What is the process for establishing an alternative cost recovery structure?

2.0 EFFECT OF WET WEATHER FLOWS

Typically, municipal utilities, wastewater authorities, or districts that are responsible for wastewater conveyance and/or treatment services have one or more of the following integrated networks of three conveyance systems within their service area:

- Combined sewer systems that convey both sanitary wastewater and stormwater inflows in the same conveyance system;
- A separate sanitary sewer system that only conveys sanitary wastewater flows; and
- An MS4 that only conveys stormwater flows.

Excessive wet weather flows that result from significant wet weather events of rainfall/precipitation due to weather patterns and climactic changes exert different levels of water quantity and quality effects, depending on the types of wastewater conveyance systems that exist within a municipality or a service territory.

2.1 Effect on Combined Sewer Systems

In service areas with a CSS, excessive wet weather flows could potentially exceed the capacity of the CSS and/or water resource recovery facilities (WRRFs). In such a situation, wet weather flows could trigger combined sewer overflows (CSOs), where the untreated combined wastewater may be discharged directly to surface receiving waters without the benefit of primary treatment. Aging and deteriorating wastewater infrastructure can further exacerbate such CSO issues. Frequent and highly publicized incidents of CSOs to rivers and streams as well as water main breaks in the nation's largest cities are the most visible manifestations of this deteriorating infrastructure problem (Gomez, 2013). Combined sewer overflows directly impair water quality, harm aquatic life, cause health hazards, and affect recreational uses of surface waters. According to the U.S. Environmental Protection Agency (U.S. EPA), CSOs are contributing sources of discharge responsible for beach closings, shell-fishing restrictions, and other waterbody impairments (https://www.epa.gov/npdes/combined-sewer-overflow-basics).

Many older large and small urban municipalities and regional facilities, including those in New York, New York; Philadelphia, Pennsylvania; Chester, Pennsylvania; Washington, D.C.; Cincinnati, Ohio (Metropolitan Sewerage District); and Kansas City, Kansas, face federal- and state-issued consent decrees or consent order agreements that require mitigation and/or elimination of CSOs and adherence to other stringent regulatory requirements. Compliance with a consent decree or consent order requires significant investments (i.e., billions of dollars) in LTCP initiatives. For example, due to consent decree mandates, the City of Atlanta, Georgia, was required to invest more than $2.0 billion over 25 years in the city's CSS to execute projects, including sewerage separation projects targeted at improving area water quality and reducing CSOs (Miller et al., 2022). What is also of importance is that most of the communities in the United States with CSO problems and issues have fewer than 10,000 people according to the U.S. EPA (2024).

2.2 Effect on Separate Sanitary Sewers

In separate sanitary sewers, especially those with aging infrastructure that may be prone to significant sewer defects and associated I/I problems, wet weather flows can exacerbate I/I problems, causing sanitary sewer overflows (SSOs). Many municipalities with SSOs also face consent orders because SSOs are also point-source discharges. Significant investments in I/I reduction and other wet weather management initiatives may be needed to mitigate SSOs.

2.3 Effect on Municipal Separate Storm Sewer System

Wet weather flows affect MS4s in several ways, including flooding, habitat degradation, stream and channel erosion, and other significant water quality issues such as sedimentation and pollution resulting from stormwater runoff.

Although the nature of wet weather flow effect varies with the type of conveyance system, a common concern among all these systems is the significant financial investment that is involved in the management of wet weather flows.

3.0 WET WEATHER CAPITAL FINANCING ALTERNATIVES

Internal and external capital program financing alternatives that are described in Chapter 4 could all be considered to effectively finance wet-weather-related capital infrastructure investments. In addition to using gray infrastructure to manage wet weather, utilities are also increasingly integrating multibenefit land-use management and private stormwater best management practices to their wet weather management portfolio and/or LTCP initiatives. For example, the Philadelphia Water Department designed the "Green City, Clean Waters" plan to mitigate CSOs and reduce water pollution (https://water.phila.gov/green-city/). This 25-year program relies on green infrastructure as part of its overall compliance strategy. Hence, additional innovative financing mechanisms are often necessary to effectively fund these types of wet weather initiatives.

The type of financing for land-management-based capital initiatives vary between public and private stormwater management initiatives.

3.1 Public Stormwater Management Capital Initiatives

Utilities can undertake recurring program initiatives such as downspout disconnection programs and residential- and neighborhood-level greening initiatives to mitigate wet weather contribution. Such programs can be budgeted as annual routine capital outlay and funded through user rates and charges.

However, other large-scale capital initiatives, including multibenefit sewer separation projects, real-time CSO controls, green infrastructure resiliency initiatives, and wetlands development projects, can be more cost effectively financed through a combination of land conservation loans, various types of applicable state and local grants, state revolving fund loans, bond financing, and other contributions. Leveraging multiple sources of funding for a large-scale project can be more cost effective than a single source of funding.

To meet expensive consent order, LTCP, and water-quality standards requirements, utilities engage in collaborative and alternative project delivery such as community-based public–private partnerships (CBP3) where practical. Alternative large-scale capital program delivery models include multientity partnerships, such as the one the District of Columbia Water and Sewer Authority (DC Water) engaged in to issue its first environmental impact bond. Another example is the CBP3 that the city of Chester, Pennsylvania, formed by bringing together Chester Water Authority, U.S. EPA, and a private partner to plan, finance, build, and maintain up to $50 million in green stormwater infrastructure over 30 years (Adaptation Clearinghouse, 2017). Such alternative and innovate delivery models could be considered, if appropriate, for accelerating wet weather initiatives and enhancing community outcomes while also mitigating the risks of financial investments, especially in implementing green infrastructure solutions.

3.2 Private Stormwater Management Capital Initiatives

Many urban municipalities also provide grants to encourage stormwater management projects for private properties. Such types of funding may require a multientity complex structure because the utility may not be able to directly administer a grant to a private property. In addition, in such situations, even if the utility provides partial funding in the form of grants, it will not be able to own and include the stormwater assets as part of its asset base. Hence, a utility may have to finance such grants as an O&M cost and not as a capital initiative.

Philadelphia, Pennsylvania, offers grant funding to private nonresidential properties under its Stormwater Grants Program, which focuses on incentivizing implementation of green infrastructure. The multimillion dollar program is outlined in the annual operating budget, and grants are primarily targeted toward CSO mitigation in combined sewer areas. Philadelphia recovers their Stormwater Grants Program related costs through sewer and stormwater rates and charges.

A key consideration in wet weather capital financing in jurisdictions with CSS is the approach used to recover the costs. In some of these jurisdictions, the urban core may have the older CSS infrastructure and the associated CSO mitigation initiatives, whereas the surrounding suburban areas may have the newer MS4 infrastructure. Therefore, utilities would have to consider these factors in evaluating alternative capital cost recovery mechanisms, and, more specifically, the appropriateness of apportioning stormwater management practices costs between sewer and stormwater utilities.

4.0 WET WEATHER COST RECOVERY

The challenge of managing wet weather flow lies not only in the diverse effect it has on conveyance systems, but also in the delineation of responsibilities among various entities. In municipalities where multiple entities are involved, wet weather financing and associated cost recovery becomes more nuanced and complex.

4.1 Delineation of Wet Weather Management Responsibilities

In some municipalities, such as in Philadelphia, Pennsylvania, the water/sewer utility is responsible for managing all aspects of wet weather flows, including LTCP/National Pollutant Discharge Elimination System requirements and MS4 requirements. However, in other municipal jurisdictions, an independent authority such as DC Water may be responsible for managing wet weather flows only in CSS and separate sanitary sewer systems, while the municipality, in this case the District of Columbia, may be directly responsible for all MS4 requirements.

Even within a MS4 service area, responsibilities are often shared by the utility and the municipality. Consequently, there are also significant differences among municipalities in the types of wet weather capital financing and cost recovery mechanisms that they use to recover O&M, debt service, pay-as-you-go, and other relevant annual wet weather costs. Table 9.1 shows examples of the diversity that exists in how wet weather management related revenue requirements are recovered from the customer base.

4.2 Key Considerations in Wet Weather Cost Recovery

Several considerations play roles in the recovery of wet weather management costs and many of these are interrelated. These are described not as stipulations, but as guidelines for defining cost-recovery policies. Municipalities and large metropolitan areas need to define their own wet weather cost-recovery policies based on practical considerations and nuances that closely meet the utility's strategic direction and needs, objectives, regulatory requirements, and other specific circumstances.

4.2.1 Equity of Cost Recovery

Cost-recovery approaches that recover all wet weather revenue requirements based entirely on sewer charges and/or taxes (see Table 9.1, Example 1) may provide for administrative simplicity and ease of customer understanding. However, such approaches may affect equity of cost recovery. Typically, sewer charges are based on the volume of water use, and taxes are based on property value, both of which have limited correlation to the magnitude of a property's wet weather contribution. Therefore, other cost-recovery mechanisms such as an impervious area-based wet weather fees could be integrated to a user-fee rate structure portfolio.

4.2.2 Recognizing Historical System Development

Equity also may be compromised by relying entirely on cost-allocation methodologies based on current use proportionality, or even current capacity

TABLE 9.1 Primary Cost-Recovery Approach for Wet Weather Revenue Requirements

	Primary approach to cost recovery		
Wet weather program	Example 1	Example 2	Example 3
CSS and separate sanitary sewer revenue requirements	Sewer charges	Sewer charges	Sewer charges and stormwater charges
MS4 revenue requirements	Sewer charges or taxes/assessments	Stormwater charges	Stormwater charges

claims, that ignore historical system development. In many cases, wastewater systems that were developed in the urban core, often with combined sewers, provided the foundation for later suburban development with separated systems. If the costs associated with these urban combined systems are allocated entirely by reference to current uses or capacity claims, separated suburban systems may be inequitably relieved of responsibility for the costs of combined systems that enabled growth in a metropolitan region. In these cases, alternative approaches can be used to recover more equitably what are fundamentally "common-to-all" costs. This can be done through wastewater charges—for example, charges based on retail customers served or flow metrics—or through different revenue-recovery mechanisms. For example, the city of Atlanta, Georgia, implemented a municipal option sales tax in part to distribute cost responsibility for their CSO consent decree across the Atlanta metropolitan region.

4.2.3 Revenue Stability

Recovery of wet weather costs through volumetric-based sewer charges can affect revenue stability because of the volatility in a customer's water use and, similarly, a tax-based revenue recovery could create revenue uncertainty because of the changing priorities of tax expenditures. Because impervious area within a property is a relatively more stable measure of stormwater contribution than the volume of water use, including an impervious area-based fee as part of a utility's overall sewer rate structure could enhance revenue stability.

4.2.4 Fee Reduction Options

The ability to provide fee-reduction options is an integral and essential component of any user-fee cost-recovery approach. Properties with large imperviousness contribute more to wet weather flows than properties with smaller impervious areas. Therefore, using alternative means such as impervious area-based wet weather fees would provide greater flexibility to afford fee-reduction options to ratepayers that deploy on-site stormwater best management practices.

4.2.5 Legal Considerations

In considering alternative approaches to wet weather cost recovery, it is critical to evaluate legal and governance considerations, including legislative authority, regulatory responsibility, potential changes to existing municipal or authority charters to assess and collect fees, and other applicable legislative and regulatory aspects.

4.2.6 Administrative Considerations

In establishing a wet weather cost-recovery mechanism, the recurring annual level of effort and associated costs involved in the various administrative aspects of managing a wet weather cost-recovery program must be considered and addressed in defining the annual revenue requirements and establishing the organizational capacity to manage such a program. Typical administrative aspects that should be considered include the following: (1) consistently managing and updating annual revenue requirements and wet weather cost delineations, (2) maintaining and updating impervious area data that may be defined as the basis of wet weather charges, (3) managing any fee-reduction program such as onsite management fee credits, (4) handling customer inquiries and appeals regarding the wet weather charge, and (5) managing the technical and process aspects of billing a parcel-based wet weather charge. Discussion about also considering a one-time implementation cost is presented in Section 5.3.

5.0 WET WEATHER RATE-SETTING APPROACH

The concept of establishing a distinct wet weather fee (often referred to as a *stormwater user fee*) is becoming more prevalent in the United States (Western Kentucky University, 2023) because of its multiple benefits discussed in the previous section. Figure 9.1 presents an illustration of the key tasks involved in developing an impervious area-based wet weather fee.

5.1 Delineation of Program Costs

To determine current and future program costs, to the extent practical, key wet-weather-related O&M activities and associated costs need to be identified. The types of O&M and capital costs that a utility includes in the delineation of wet weather program costs depend on a utility's overall mission, operational level of service, community needs, regulatory obligations, and system infrastructure.

In jurisdictions that only have MS4 service areas, typical stormwater program cost elements include stormwater collection and drainage, green infrastructure best management practices (BMPs), asset mapping and management, stormwater treatment, MS4 permit compliance and enforcement activities, source water and/or watershed protection programs, and pertinent planning and administrative activities.

In jurisdictions that include a CSS, the complexity of stormwater cost delineation increases (Water Environment Federation, 2013) because stormwater is also conveyed through the CSS to WRRFs. In service areas with a

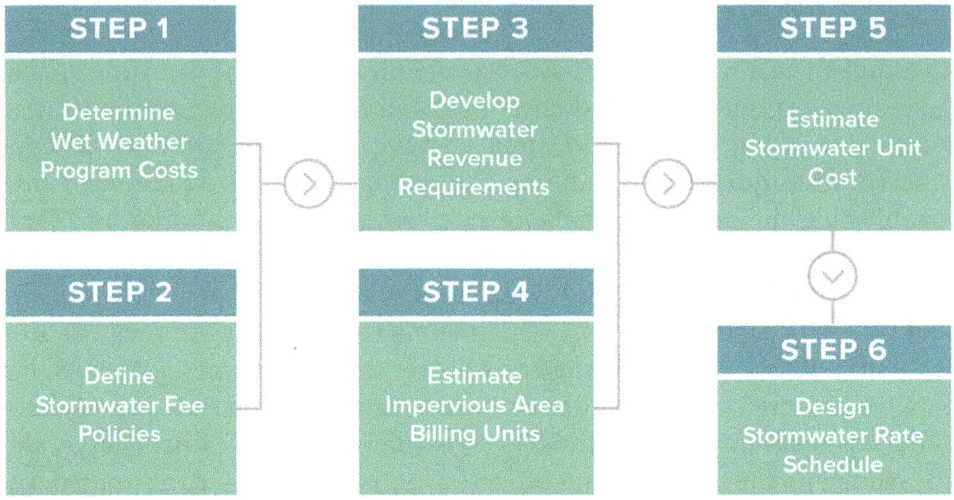

FIGURE 9.1 Key Steps in Developing a Wet Weather (Stormwater) Fee

CSS, utilities incur significant O&M and capital costs to mitigate CSOs and meet consent decree requirements where applicable. Therefore, to determine total wet weather costs, utilities could allocate a reasonable portion of the CSS O&M and capital costs to stormwater services.

The CSS costs that could be allocated include conveyance, pumping, treatment, CSO-mitigating green initiatives, public education, and other pertinent administrative costs. To be defensible, utilities must exercise sound engineering science-based principles and use available budget and actual cost data to allocate a portion of the CSS costs to stormwater. For example, O&M costs relating to inlets/catch basin cleaning and debris disposal in CSS areas could be allocated 100% to stormwater. All other CSS costs could be apportioned between sanitary sewer service and stormwater service.

5.2 Stormwater Fee Policies and Legal Considerations

In establishing a distinct stormwater or wet weather fee, it is prudent to first develop a set of key policies such that key assumptions used in the cost-of-service analysis, fee methodology, and rate design are consistent and defensible. The framework should clearly define the basis for key aspects, including the following:

- Apportioning of CSS and MS4 costs to stormwater-revenue requirements,
- Geographical considerations,

- Capital program financing and other financial policies pertinent to stormwater-revenue requirements,
- Stormwater customer classification,
- Impervious area estimation methodology,
- Billing and enforcement,
- Rate structure design and stormwater credits, and
- Other pertinent considerations.

In addition to the development of a policy framework, early in the process, utilities must conduct a due diligence review of federal, state, and local enabling legislation, the charter under which the utility operates, to affirm that the utility has the authority to implement a stormwater or wet weather fee within its jurisdiction. It is also important to understand state and local statutes and ordinances regarding the ability to assess stormwater fees on certain classes of properties and user-fee voter approval/referendum procedures.

5.3 Stormwater Revenue Requirements

The next critical step is to develop the stormwater-revenue requirement that is to be recovered from a separate stormwater user fee or a wet weather fee. Once stormwater program costs are delineated, as discussed in Section 5.1, a comprehensive multiyear revenue-requirement analysis should be performed. Typically, annual revenue requirements include recurring annual O&M costs, debt service expenses, pay-as-you-go cash financing, reasonable operating capital, and any equipment reserve requirements and debt service coverage requirements, where applicable.

When establishing a new stormwater user fee, it is also important to consider the inclusion of one-time implementation costs and any one-time program ramp-up costs in the determination of revenue requirements.

Potential annual revenues from other sources such as stormwater plan and inspection reviews, any stormwater low impact development in-lieu fees, and impact fees should be evaluated and deducted from the total annual revenue requirements to determine the net stormwater-revenue requirement that is to be recovered from stormwater or wet weather user fees.

5.4 Impervious Area Billing Units

A critical challenge in establishing a wet weather fee or a stormwater user fee is defining the basis for assessing the fee. Because stormwater contribution from a property cannot be directly measured, other surrogate measures such as impervious area are often used to develop a fair approximation of each property's stormwater demand on a utility's wastewater system. Multiple cost-causative factors such as a property's topography, pollutant contribution, and volume and rate of runoff can influence a property's stormwater demand. However, determining billing units based on these multiple parameters is often not practical because it is technically and administratively complex. Hence, the use of impervious area as a basis for determining billing units is most widely used and generally viewed as a best practice (Black & Veatch, 2021).

Various data sources, including a municipality's tax assessment system, geographic information system (GIS), aerial and orthographic imagery, and infrared imagery, can be effectively used to estimate impervious area. Typically, properties are divided into stormwater classes such as single-family residential, commercial, institutional, parking, parks, and so on based on the granularity of land-use information that is readily available in tax assessment and GIS systems.

Subject to data availability and the costs involved in developing the impervious area, a combination of approaches can be used to determine the impervious area for each stormwater class. The aggregate of the impervious area estimated for all classes is then defined as the systemwide impervious area.

With respect to wet weather fees, customers can seek fee reduction in the form of stormwater credits for any fully functional onsite stormwater-management practices in place. In addition, customers may also seek fee adjustments because of potential data inaccuracies or exceptions in the impervious area. Therefore, to account for potential revenue reductions because of stormwater credits and appeals, it is imperative that the initial estimate of systemwide impervious area be reduced before finalizing the total billable impervious area.

The billing unit of measure for a stormwater or wet weather fee is often expressed using a single parameter such as 100 or 500 sq ft (9 to 46 m^2) of impervious area or an equivalent runoff unit (ERU), which is also referred to as an *equivalent* residential *unit*. The average or median impervious area square footage of the single-family residential class is defined as one ERU. The total systemwide billable units are then expressed in terms of increments of 500 sq ft of impervious area or ERUs.

5.5 Stormwater Unit Rate

Just as water and sewer rates are often expressed in terms of cost per cubic meters (or hundred cubic feet or thousand gallons; 1 Ccf = 2.83 m^3), a stormwater user rate is expressed as rate per 100 or 500 sq ft (1 sq ft = 0.09 m^2) of impervious area or rate per ERU ($/ERU). Depending on the billing unit of measure that is defined, the annual net revenue requirement is divided by the total estimated billable stormwater units to determine the systemwide monthly or annual stormwater unit rate. In the following example, 1 ERU is 1500 sq ft, the annual stormwater revenue requirement is \$8,615,000, and the total system impervious area is 217 695 000 sq ft. The following equations provide example illustrations of the calculation of a stormwater unit rate for a representative year, which then provides a defensible basis for designing the stormwater user-fee rate structure:

$$\begin{gathered}\text{Total system ERUs} = \text{Total system impervious area/1500} \\ \text{Total system ERUs} = \text{217 695 000 sq ft/1500 sq ft} = \\ \text{145 130 sq ft}\end{gathered} \tag{9.1}$$

$$\begin{gathered}\text{Annual stormwater ERU rate} = \text{Annual stormwater} \\ \text{revenue requirement/Total system ERUs} \\ \text{Annual stormwater ERU rate} = \\ \text{\$8,615,000/145 130 sq ft} = \text{\$59.36/sq ft}\end{gathered} \tag{9.2}$$

$$\begin{gathered}\text{Monthly stormwater ERU rate} = \text{Annual stormwater} \\ \text{ERU rate/12 months} \\ \text{Monthly stormwater ERU rate} = \text{\$59.36/sq ft} \div 12 = \\ \text{\$4.95/sq ft per month}\end{gathered} \tag{9.3}$$

Based on the ERU rate of \$4.95, in the above example, a monthly stormwater charge for a single-family residential property with 3000 sq ft of impervious area would be calculated as follows:

$$\text{3000 sq ft/1500 sq ft} = \text{2 ERUs}$$

$$\text{2 ERUs} \times \text{\$4.95/sq ft per month} = \text{\$9.90/sq ft per month}$$

5.6 Stormwater Rate Structure Design

Most of the rate-policy objectives discussed in Chapter 8, such as fairness and equity, administrative ease, customer understanding, and affordability considerations, are applicable in the design of a stormwater or wet weather fee structure. Typically, key components of a stormwater rate structure include the following:

- Residential. Single-family residential properties (often defined as buildings with up to three or four dwelling units) can be charged based on one of the following three alternatives: (1) a uniform monthly fee where all properties pay the same fee; (2) a tiered impervious area fee where, based on a distribution analysis of the impervious area, three to five tiers of impervious area are designed; or (3) an individual calculation for each parcel in which the monthly ERU rate is applied to each parcel's specific impervious area. Each approach has its benefits and disadvantages that have to be carefully considered in selecting the best-suited alternative.

- Nonresidential. Because of the significant differences in impervious area characteristics among various nonresidential classes of properties, including multifamily properties and condominiums, it is prudent to calculate the stormwater charge for each parcel by applying the monthly ERU rate to each parcel's specific impervious area.
- Minimum charge. Many utilities also establish a minimum charge that either equates to one ERU or the monthly fee of the lowest impervious area tier if the utility has a tiered rate structure.

6.0 CASE STUDIES

6.1 District of Columbia

6.1.1 Overview

Stormwater management responsibilities are shared by two entities in the District of Columbia. The District Department of the Environment (DDOE) is responsible for managing the separate stormwater system and compliance with the District's MS4 permit. DC Water is responsible for the CSO LTCP and the management of wet weather issues within combined sewer areas. Each agency has a distinct wet weather fee for recovery of wet-weather-related costs. This approach to wet weather cost recovery aligns with Example 3 in Table 9.1.

6.1.2 Key Driver

Approximately 65% of the service area is a separate stormwater system and 35% is a CSS. Originally, DC Water collected all fees; however, in 2007, DDOE was established and the MS4 compliance duties were transferred to DDOE. Both DDOE and DC Water strive for equitable recovery of wet-weather-related costs.

6.1.3 Costs Recovered

The DDOE has a "stormwater fee" that recovers costs associated with regulatory compliance related to the MS4 permit, including green infrastructure retrofits. In addition, DC Water has a clean rivers impervious area charge (CRIAC), which recovers CSO consent decree compliance costs, including operational upgrades, capital investments, and debt service.

6.1.4 Rate Structure

Both the DDOE stormwater fee and the DC Water CRIAC are based on actual impervious area calculations per parcel and both agencies bill their fees using an ERU that is based on the average amount of impervious surface on residential properties. In addition, both entities offer a credit program for their customers to earn a discount on their fees. The DDOE also offers a stormwater retention credit trading program.

The District of Columbia is unique in that there are two distinct wet weather fees assessed by two different agencies. This requires the need for a clear definition of roles, responsibilities, and funding sources for the activities required to enhance stormwater management.

6.2 Philadelphia, Pennsylvania

6.2.1 Overview

Philadelphia Water Department (PWD) is responsible for retail water, wastewater, and stormwater services within the city's jurisdiction. Approximately 40% of the service area is a separate stormwater system and 60% is a CSS. The city recovers wet weather costs through a sanitary sewer charge and a stormwater management service (SWMS) charge. This approach to wet weather cost recovery aligns with Example 3 in Table 9.1.

6.2.2 Key Driver

A key driver for stormwater management costs is the implementation of the "Green City, Clean Waters" initiative. This plan is an integral part of the city's CSO LTCP and its holistic approach to managing water resources for the city. The 25-year plan will transform the health of the city's creeks and rivers primarily through a land-based approach of implementing green stormwater infrastructure projects.

6.2.3 Costs Recovered

The SWMS charge is designed to recover all of the wet weather costs associated with managing the MS4 system and portions of the CSS system, including LTCP requirements. In addition, O&M of PWD's inlet and catch basin cleaning, stream restoration, and a portion of the green infrastructure program and administration program costs are also recovered.

6.2.4 Rate Structure

The city's SWMS charge is based on two parameters: the average gross area square footage and the average impervious area square footage; it also includes a monthly billing and collection charge. In addition, PWD offers stormwater credits and incentives programs, the costs of which are proportionally funded through wastewater and stormwater rates.

When the SWMS charge was created in July 2010, affordability was a key concern for business and nonprofit customers because, under the new system, these nonresidential customers pay based on their impervious "footprint." To mitigate this effect, the PWD phased in the new charge over 4 years and established a rate increase cap for nonresidential customers.

7.0 SUMMARY

The characteristics of wet weather and the extent of wet weather effects vary significantly among municipalities and watersheds. Determining wet weather costs and establishing an equitable approach to the recovery of those costs can provide for revenue stability, equity of cost recovery, flexibility to encourage public and private stormwater BMPs, and the ability to better align user fees with program needs, costs, and customer benefits.

However, determining alternative wet weather financing mechanisms such as public–private partnerships and/or establishing a separate wet weather fee are critical policy decisions. These decisions need to be made based on a careful and diligent evaluation of myriad aspects, including delineation of service responsibilities, inter- and intra-governmental contractual agreements, program needs and costs, the economic environment, customer demographics and affordability, legislative feasibility, and administrative capacity and costs.

8.0 REFERENCES

Adaptation Clearinghouse. (2017, June). Chester, Pennsylvania green stormwater infrastructure plan and community-based public–private partnership. https://www.adaptationclearinghouse.org/resources/chester-pennsylvania-green-stormwater-infrastructure-plan-and-community-based-public-private-partnership.html

Black & Veatch. (2021). 2021 *Stormwater utility survey: A Black & Veatch report.*

Environmental Finance Advisory Board. (2020). *Evaluating stormwater infrastructure funding and financing—A report from the Stormwater Infrastructure Finance Task Force Work Group of the Environmental Finance Advisory Board.*

Gomez, A. (2013). Water infrastructure—approaches and issues for financing drinking water and wastewater infrastructure. Testimony before a Congressional Subcommittee, United States Government Accountability Office (GAO-13-451T).

Miller, A., Ebelt, S., & Levy, K. (2022). *Combined sewer overflows and gastrointestinal illness in Atlanta, 2002-2013: Evaluating the impact of infrastructure improvements.* https://pubmed.ncbi.nlm.nih.gov/35580035/

U.S. Environmental Protection Agency. (2024, May). *National Pollutant Discharge Elimination System (NPDES), Combined sewer overflow frequent questions.* https://www.epa.gov/npdes/combined-sewer-overflow-frequent-questions

U.S. Environmental Protection Agency. (2024, September). *National Pollutant Discharge Elimination System (NPDES), Combined sewer overflow basics.* https://www.epa.gov/npdes/combined-sewer-overflow-basics

Water Environment Federation. (2013). *User fee-funded stormwater programs* (2nd ed.).

Western Kentucky University. (2023). Stormwater utility survey.

10

System Development Charges

1.0 INTRODUCTION

A comprehensive financial plan is critical to ensuring that a wastewater utility is adequately prepared for ongoing system operation and replacement requirements and future capacity expansion needs. This plan should consider the full array of capital-funding alternatives available to the utility, as discussed in Chapter 4. One possible funding source for capital projects is system development charges (SDCs).

System development charges have been used since the 1920s, as evidenced by the U.S. Department of Commerce's Standard Planning Enabling Act, which was enacted in 1922. Other common names for SDCs include, but are not limited to, *development fees*, *impact fees*, *capacity fees*, and *capital recovery fees*. System development charges are one-time charges to customers when they connect to the system or by developers as part of the permitting or planning process. The SDCs are also charged to existing connections requiring increased capacity. Typically, system development charge proceeds are used to pay for capital projects related to growth and/or reimburse existing customers for past system capacity investments.

Capital projects required to meet the demands of growth are often a burden on existing wastewater ratepayers. Through the use of SDCs, costs associated with growth may largely be shifted to new customers and away from existing wastewater customers over time, helping achieve intergenerational equity among customers. Application of these fees assists the utility in implementing a "growth pays for growth" policy and helps with future utility capital improvement planning.

Typically, SDCs are used to pay for backbone water resource recovery facilities (WRRFs), including collector mains, interceptor mains, outfall sewers, and lift stations. However, where applicable and allowable, costs associated with nonbackbone facilities, equipment, minimum reserve requirements, and fee administration activities may be recovered through SDCs.

This chapter presents legal and regulatory considerations that contributed to the development of standards in formulating and implementing SDCs, the steps involved in developing and implementing SDCs and the various methodologies and concepts, and SDC program management issues, including the application of revenues.

2.0 LEGAL CONSIDERATIONS

Before implementing SDCs, the utility should develop its philosophy and capital improvement plan (CIP) for further development and maintenance of its infrastructure. It should also review pertinent legislation, state statutes, local municipal codes, and judicial rulings related to SDCs. Utilities need to be aware of the legislative authority within the state in which they operate in developing such fees. Typically, authority for charging SDCs comes from (1) specific enabling legislation; (2) general home-rule powers, which provide local governments the authority to establish fees and charges for local government facilities; (3) broad police power to protect general health, safety, and welfare of the community through provision of services; and (4) utility rate-setting authority. Laws regarding SDCs can often be vague, and misapplication of concepts and approaches may lead to legal challenges. Utilities should seek the advice of legal counsel familiar with SDCs, especially when new SDCs are being considered for implementation.

Legal frameworks established in the early 1900s granted local government to regulate growth, giving rise to SDCs. Since that time, state and local legislation and court decisions have addressed SDCs and, in that process, have established various guidelines and standards for establishing these charges. System development charges have increasingly become an accepted revenue source in supporting growth. Although SDC statutes and judicial rulings vary by state, prevailing U.S. Supreme Court rulings can provide relevant guidelines.

2.1 Right of Local Governments to Regulate Development

State and local governments have the right, through the 10th Amendment of the Constitution, to regulate land development. The power of regulation rests with states that elect to delegate that authority to local governments for purposes of guiding land developments (Porter, 1997). Opponents of SDCs have argued that these fees were not authorized by state statute. To address this issue, many states have adopted legislation officially authorizing the use of SDCs. Some of these statutes authorize very specific uses of SDCs, while others have adopted more general authorizing statutes (Nicholas et al., 1991).

2.2 Regulation Versus Taxation

In the past, SDCs have repeatedly been challenged on the basis that they constitute taxes rather than fees for service. It is now generally, but not totally, accepted that SDCs are user charges rather than taxes. This is based on the rationale that the fee is voluntary and benefits the paying entity based on the use of the facility. System development charges for new wastewater users are thus fees distinguishable from taxes because they are related to cost of construction and use of the facilities and are proportionately charged to users who benefit from facilities.

2.3 Banberry Factors (Standards of Reasonableness)

One of the most influential court cases in the history of development fees was *Banberry Development Corporation v. South Jordan City* (1981). In this case, the Utah Supreme Court held that, in establishing development fees, local governments must consider the share of capital costs to be borne by newly developed properties relative to the costs already borne by existing properties. Specifically, the court identified the following seven factors that an entity must consider:

1. Cost of existing capital facilities;
2. The manner of financing existing capital facilities;
3. The relative extent to which newly developed properties and other properties in the municipality have already contributed to the cost of existing capital facilities;
4. The relative extent to which newly developed properties and other properties in the municipality will contribute to the cost of existing capital facilities in the future;
5. The extent to which newly developed properties are entitled to a credit because the municipality is requiring their developers or owners to provide common facilities that have been provided by the municipality and financed through general taxation or other means in other parts of the municipality;
6. Extraordinary costs, if any, in servicing newly developed properties; and
7. The time–price differential inherent in fair comparisons of amounts paid at different times (*Banberry Development Corporation v. South Jordan City*, 1981).

The Banberry case established procedural and substantive guidelines for cases where SDCs are challenged and provided guidance for policymakers in establishing an equitable program (*Banberry Development Corporation v. South Jordan City*, 1981).

2.4 Rational Nexus

Rational nexus refers to the concept that there needs to be a reasonable connection between the following:

- The new development that will pay the fees and the need for facilities;
- Growth needs and levels of cost to meet that need compared to the cost to serve others;

- Identified costs and the fee level;
- Identified costs and the amount of revenue generated by the fee; and
- The cost to the utility of new development and the amount of the fee collected.

This test is referenced in many court cases and provides the guidelines found in state statutes for SDCs.

2.5 Good Faith Intent

In 1997, Arizona courts found that a municipality needs only to develop a plan that demonstrates "good faith" intent to use SDCs to provide growth-related services within a reasonable time. This case suggests that the SDC programs are not required to be precise, but that they do have to be formulated based on sensible planning.

3.0 PLANNING A SYSTEM DEVELOPMENT CHARGE PROGRAM

Local governments across the country have varying policies related to growth. At one extreme are communities actively pursuing economic development, while others seek to limit or control further growth and development. Although SDCs should not be used as a disincentive for growth, this type of fee is an important tool in ensuring adequate infrastructure to serve growth.

3.1 Capital Program Planning

Capital and financial planning helps to ensure that SDCs are legally defensible. Planning begins with a capital planning process, as noted in Chapter 4, Table 4.1, that determines existing and future system capacity needs and specific capital projects required to meet those needs. It takes knowledge, time, and effort to create legally sound and politically feasible fee programs. Typical actions required by a utility to implement system development charges include the following:

- Confirming that the local government has authority to establish such fees by statute or other means;
- Adopting a facility plan, masterplan, or CIP that projects service area growth, assesses existing system capacity, identifies additional capacity created by future projects, outlines projects necessary for growth, and identifies funding sources for projects;

- Developing a fee structure or method consistent with legal requirements; and
- Monitoring programs to ensure that revenues benefit the intended growth area.

Utilities facing minimal growth may require only minor modifications to the wastewater system for each additional customer. However, utilities experiencing rapid growth may require significant capital projects to serve planned development. In the latter case, it is particularly important that the utility adopt a CIP (often part of a master plan or other system infrastructure plan) that includes the following:

- Projected development throughout the planning period;
- Distribution of growth throughout the service area;
- Capacity requirements of growth, in terms of flow and loadings;
- Existing system loadings and facility capacities;
- List of planned capital improvements to address various needs (replacement, rehabilitation, expansion, etc.);
- Designated/identified funding sources for each project; and
- Estimated timeframe for completion of capital improvements.

Capital planning may be explicitly addressed in master plans, or it may be separate from such plans. Capital improvement plans may address the need for creating new WRRF capacity and needed improvements to existing facilities to meet designated service demands or regulatory requirements. The CIP used to develop the SDCs should identify the costs of growth-related facilities separate from the improvement and regulatory-related facility costs for existing customers.

Projected capacity requirements are based on growth assumptions applied to area maps and land-use assumptions such as residential, multifamily, commercial, or industrial uses. New capacity requirements can also be based on existing development in the area to be served. Once land-use requirements are categorized and the needs of existing development have been estimated, the utility can assess additional capacity requirements. Furthermore, the CIP should forecast timing of development, enabling the sizing and timing of WRRFs.

The use of engineers who specialize in utility master planning is important in the development of detailed capital projects, schedules, cost estimates, and project prioritization. Once a reasonable plan has been developed by the utility, it may be advantageous to create a public forum, such as citizen committees, to provide comments on CIP priorities and alternatives. Because growth-related projects can place a financial burden on existing customers (if fully funded through user rates), utilities often look for other means of funding these projects through direct charges to the benefitting users.

3.2 Level of Service

An important consideration in the development of SDCs is the level-of-service (LOS) standard. When SDCs are designed, wastewater demands are compared to the typical demand of a single-family residential unit to establish an equivalent residential unit (ERU) or equivalent residential connection (ERC) because, typically, they are the most common types of service and demands are more consistent within the class of customer. The LOS is an indicator of the service capacity provided by the wastewater system to a customer and is expressed in terms of capacity allocable to an ERU or ERC per unit of demand (e.g., gallons or hundred cubic feet [m^3]) on an average or peak basis.

Typically, LOC information is established in capital-planning documents such as facility or master plans, local code of ordinances, and state agencies, among others. However, LOS standards should be evaluated and justification for the LOS should be provided when developing SDCs. For example, given that LOS standards are user-driven, a review of flows will be able to provide information as to whether a LOS standard is reasonable.

3.3 System Development Charge Calculation

In calculating SDCs, a utility needs to select a fee structure and develop a fee schedule based on selected units of service. Methodological approaches should be evaluated and selected with careful consideration of state regulations, data availability, and local growth policies. Ultimately, there are several principles the utility needs to satisfy in developing the methodology; these are as follows:

- Fee is proportionate to proposed effect,
- Fee proceeds are used to provide infrastructure serving the growth area (which may be the entire service area),
- Fee methodology is uniform and consistent,
- Other sources of funding are considered for capital improvements, and
- Fee includes only eligible growth-related costs.

Specific approaches to developing the fee structure and schedule are discussed in the following section.

4.0 SYSTEM DEVELOPMENT CHARGE CALCULATION METHODS

4.1 Examples of System Development Charge Methodologies

There are three broadly recognized structures of system development charges (Galardi et al., 2004), which are as follows:

1. System buy-in approach based on existing facilities;
2. Planned facility/growth approach based on the projection of capacity-enhancing system improvements; and
3. Combined approach, which considers both existing and planned future facilities.

Each option is discussed in detail in the following sections. It is important to determine the underlying philosophy before adopting a specific methodology and check applicable state statutes and case laws for permissible methodologies.

4.1.1 Buy-In Approach

Under this approach, new customers are required to "buy-in" to existing system facilities, generally at a rate that reflects the previous investment of existing customers per unit of total capacity (capacity buy-in). A buy-in-type approach is easy to administer and is most appropriate where current system facilities have adequate capacity to serve both existing and future customers, the forecast of future system investment is minimal, and where existing facilities are not scheduled for replacement in the near future (American Water Works Association, 2017).

The rationale behind the capacity buy-in approach is that new customers should be charged for existing available capacity at a rate consistent with the average value of available capacity of the existing system (see the System Valuation section for considerations related to establishing appropriate system value—the numerator of the buy-in fee equation). To the extent that there is sufficient available capacity in the existing system to serve growth, typically, the capacity buy-in approach is a reasonable basis for determining growth-related costs. However, if the existing system has little available capacity and the cost of providing new capacity (on a per-unit basis) is higher than existing facilities (e.g., because of higher standards and fewer grants), then a capacity buy-in approach may not generate sufficient revenues to fully fund the total capacity needs of growth.

4.1.1.1 System Valuation

A key methodological decision under a buy-in fee methodology is how to value the utility's system assets. The valuation method selected depends on the individual system and the objective of the utility managers. Valuation approaches include the following:

- Original cost,
- Original cost less depreciation (OCLD),
- Replacement cost, and
- Replacement cost less depreciation (RCLD).

Original cost recognizes the value of the system constructed by existing customers at the time the assets were constructed. Depending on when the assets were constructed, original cost may not be a completely accurate representation of the current value of the assets. On the one hand, assets deteriorate over time and will need to be replaced eventually. On the other hand, the cost of replacing assets at current construction costs will likely be more expensive than when the original assets are constructed. Assets constructed many years before the buy-in calculation is conducted may not be representative of the value of current value of those assets.

Original cost less depreciation recognizes the deterioration of assets over time by excluding accumulated depreciation based on the useful lives of the assets. If, for example, an asset with a useful life of 50 years was constructed 25 years ago, the OCLD of that asset would only be 50% of its original cost. While OCLD provides recognition of asset depreciation over time, assets that are at or near the end of their useful lives should be closely examined to ensure that they are being valued appropriately. Because depreciation is an accounting evaluation, typically made based on categories of assets in aggregate, it is possible that some individual assets may be fully depreciated but still in-service and available to provide capacity to new customers. While both original cost and OCLD relate to what was actually paid for the asset by existing customers, the value is in nominal dollars and does not recognize that, absent the existing capacity in existing assets, new assets would need to be constructed to serve new customers at a higher cost.

Replacement cost values system assets at the cost to replace them today, regardless of when they were originally constructed. This approach recognizes the benefit of existing capacity to new customers based on the alternative:

constructing this same capacity today, in the form of a new asset, at today's costs. Replacement cost is determined by inflating or escalating each asset's original cost based on an inflation index. Inflation indices assign a numeric value to each year that increases over time. The increase in that value is based on the increased cost of a representative group of goods and services. To determine replacement cost, construction cost indices—which track the cost of skilled labor and construction materials—can provide an appropriate basis for determining replacement cost. To determine replacement cost, each asset or group of assets would be escalated based on the index value today compared to the value when the asset was constructed. If, for example, the index value has increased 50% from the date of construction to today, the asset's original cost would be increased by 50%. Care must be taken to select a construction cost index whose representative goods and services are relevant to the assets being valued.

Replacement cost less depreciation includes accumulated depreciation in recognition that existing assets have deteriorated over time and will need to be replaced eventually. Similar to OCLD, RCLD recognizes that new customers are not buying into a brand-new system. The RCLD can be calculated in the same fashion as OCLD. If an asset is 50% depreciated, the replacement cost of that asset would be reduced by 50% to determine its RCLD. Replacement cost and RCLD, while disconnected from the original cost incurred for system assets, can provide an accurate representation of their value to new customers today.

Another important consideration under the buy-in approach is the method used to finance system assets included in system value. System assets may be financed using a variety of approaches, including

- Debt financing,
- Contributions in aid of construction,
- Grants, and
- Ad valorem tax payments.

Knowledge of relevant enabling legislation and case law is necessary to determine which of the aforementioned factors may be legally required versus simply being consistent with equity objectives.

Debt financing of growth-related system assets can improve alignment between the cost effect of growth-related projects (over time via principal payments) and the recovery of those costs from new development (over time as SDCs are paid). Because, typically, the timing of SDC revenues is not aligned with principal and interest payments, it is typical to recover the cost of principal and interest payments in customer rates. To avoid double charging, first, through SDCs and, again, through general rates and charges for wastewater service that are used to retire the debt, it is important to account for any rate recovery for principal payments when determining and applying SDCs.

A common approach to addressing debt principal involves excluding the value of outstanding principal used to fund growth-related system assets. This approach accounts for principal repayment through rates when determining the value of system assets to be included in the fee. An alternative approach involves including the full value of system assets (including debt-funded assets) but providing a credit for the portion of growth-related debt principal that will be repaid through rates. Typically, the credit is equal to the estimated present value of future rate contributions toward the debt principal. This approach is more complex because the portion of principal repayment through rates must be estimated based on projections of rate revenue and the proportion of that revenue that is related to principal repayment. Additionally, the amount of the credit will decrease as the outstanding principal of debt decreases. Therefore, the credit will need to be adjusted regularly to reflect the cost of capacity accurately. However, a credit approach may address potential equity issues with new development because the level of credit is often tied to when a development connects to the system.

Another consideration when debt-financing growth-related system assets involves the treatment of interest payments. Because growth-related systems assets are financed and constructed up front and recovered via SDCs over time, existing customers carry the cost of improvements for future customers until they connect. Accordingly, it is appropriate to consider interest payments related to financing growth-related system assets when valuing the system for purposes of calculating buy-in fees.

When establishing the value of system assets, external sources of financing such as developer contributions in aid of construction, ad valorem tax funding, and grant funding should be reviewed. Developer contributions of growth-related assets allow the planning and development of a specific area without the financial and construction constraints of the utility. The use of ad valorem taxes to finance growth-related system assets provides recognition that the availability of utility service to a new service area may generate additional tax revenues. Typically, these two sources are excluded from the valuation of system assets to avoid double-recovery of these costs. State and federal grants are other potential external sources of funding. While these may not present a double-recovery issue, typically, they are excluded from system value. Grants offset costs that would have been incurred by the utility and its existing customers, so there is no need to recover these costs from developers.

Table 10.1 illustrates determination of system value under a buy-in approach. In the example provided, assets are valued based on the net book

TABLE 10.1 Example of System Valuation

	Net book value
Water resource recovery facility	$93,533,000
Pumping and lift stations	$4,092,000
Collection system	$79,723,000
Solids processing and disposal system	$3,986,000
Existing system value	**$181,334,000**
Less (offsets)	
Debt service (outstanding principal)	($40,000,000)
Federal funding (grants)	($1,000,000)
Contributions	(42,630,000)
System valuation for buy-in fee	**$97,704,000**

value (original cost less depreciation) approach, and deductions include outstanding debt principal, federal funding, and developer contributions.

4.1.1.2 Capacity Determination

The next component in calculating system development charges under a buy-in approach is the determination of system capacity. The appropriate capacity measure under the capacity buy-in method is *total system* capacity (as opposed to "used" capacity for the equity buy-in approach). In either case, capacity may be stated in terms of hydraulic or loading capacity, or in terms of ERUs served. Equivalent residential units are the number of units in the system of varying size expressed in terms of a common unit (typically a residential dwelling). In this case, the remaining facilities (such as multifamily, commercial, institutional, and industrial) are assigned multiple equivalent units in proportion to their total contribution to capacity, relative to that of a single-family dwelling unit.

Typically, total system capacity is determined based on facility-sizing criteria and wastewater permit requirements. Existing used capacity can be determined from WRRF records and billing data. The systemwide unit cost is calculated by dividing the system valuation by the selected capacity measure. Table 10.2 shows these sample calculations under the capacity buy-in approach.

4.1.2 Planned Facility/Growth Approach

The planned facility/growth approach is based on the principle that new system users should be responsible for the cost of the latest or next increments

TABLE 10.2 Example of Capacity Buy-In Method

Existing system valuation	$97,704,000
Total capacity in ERUs	170 000
Average cost per ERU	$575

of capacity that they cause to be constructed. This fee recovers growth's share of planned additions to the system. Typically, a utility relies on its CIP to estimate cost and capacities of growth-related projects. The capacity resulting from additional facilities will be used in the fee calculation. Selecting the appropriate capacity can be determined by (1) using total new WRRF capacity or (2) capacity of new projects weighted by individual project costs (Corssmit et al., 2002). System development charges may be phased in as development in a growth area progresses and capacity use increases by using an inflationary pricing structure.

Utilities should avoid including overlapping cost in both the SDCs and wastewater user charges. As in the buy-in approach, capital contributions and grants for additional facilities should not be included in calculation of the fee. The objective of the planned facility method is that system expansion needed to serve new development can be accomplished with limited effect to existing wastewater user rates. This method is appropriate when all or a significant portion of the CIP serves growth and available facilities cannot accommodate growth. Table 10.3 provides an illustration of a planned facility approach. In this case, individual projects are analyzed to determine the portion of costs associated with system expansion and capacity versus rehabilitation or replacement.

TABLE 10.3 Example of Planned Facility/Growth Approach

Capital improvement plan	Project budget	Projects attributable to system expansion and growth	Growth costs
Water resource recovery facility improvements	$8,000,000	50%	$4,000,000
Pipe replacement	$6,000,000	0%	$0
New pumping station	$1,000,000	100%	$1,000,000
Cost of growth-related projects			**$5,000,000**
Total new system equivalent residential units (ERUs)			10 000
Unit cost per ERU			$500

4.1.3 Combined Approach

Increasingly, in response to the stated goal to charge new customers for the full cost of growth and thereby avoid the subsidization of new customers by existing customers, many state laws allow utilities to implement both approaches described previously into a single combined approach. Typically, this combined approach applies when current system facilities could serve future customers and a portion of the CIP is also related to growth. The combined approach includes the following two separate elements:

1. System reimbursement component. Includes a portion for the new customer to pay for an equitable share of existing facilities.
2. Incremental new capacity component (also referred to as *growth-related improvement component*). Includes future facilities that will be constructed to accommodate growth.

Typically, this approach is the most technically rigorous of the system development charge calculation approaches. It involves explicit determination of available capacity value in the existing system and apportionment of future capital costs between existing users and new development. The reimbursement fee component is determined by dividing the value of available capacity in the existing system by estimated growth units during the planning period. The improvement fee component is determined by dividing the value of future capacity-increasing costs by estimated growth units. So, unlike the planned facility approach described previously, which only recovers the future capacity costs related to serving new growth, the combined approach also recovers the costs of available capacity of the existing system.

Some states, such as Oregon, explicitly allow for use of a combined methodology. Other states have case law that supports this approach. For example, a 2001 Colorado Supreme Court decision found that an SDC methodology including a reimbursement component and a growth-related improvement component was justifiable and defensible. A defensible method requires rational policy, application of appropriate laws, collection and analysis of relevant data, sound asset valuation, and cost-allocation methodology. The Colorado ruling has since been adopted by several other states and its implications may be useful to utilities elsewhere (Corssmit et al., 2002).

5.0 SYSTEM DEVELOPMENT CHARGE SCHEDULE STRUCTURES

The fee structure is the mechanism for determining the costs to be recovered from new development. Of equal concern to local governments and the development community alike is the way fees are assessed to different types, sizes, and locations of development. The applicable SDC for a specific development is determined by multiplying the systemwide unit cost (as defined by one of the methods described previously) by the estimated capacity requirements (Galardi et al., 2004).

5.1 Scaling Measures

At the very least, typically, the fee schedule provides one or more scaling measures for assessment of development fees to different types and sizes of developments. The use of scaling measures in calculating development fees is designed to ensure that customer bases that are larger or use infrastructure systems more intensively pay the associated costs of capacity required to serve them. There are several measures used in the wastewater industry to represent use of capacity. These are as follows:

- Wastewater demand measured or estimated by appropriate flow and strength parameters,
- Water meter size,
- New plumbing fixtures,
- Dwelling unit count, and
- Square footage.

In developing or choosing a scaling measure for wastewater SDCs, the choice of an indicator must be easy to explain to the public, defensible, and have data supporting how the measure was derived. Estimated wastewater demand measured by appropriate flow and strength parameters is, in theory, the most equitable indicator; however, it can be difficult to develop and administer. Typically, wastewater strength is measured using biochemical oxygen demand (BOD) and total suspended solids. In some cases, chemical oxygen demand (COD) may be used in place of BOD. Other measurements that may be used include total Kjeldahl nitrogen and phosphorus. The most common indicators used in wastewater to represent capacity use are new plumbing fixtures, building square footage, and water meter size because of simplicity, ease of understanding, and ease in fee administration. An example of an SDC schedule based on meter size and the unit cost of capacity from Table 10.3 is illustrated in Table 10.4.

The utility should assess which indicator best apportions capacity cost to customers based on land use and characteristics of the service area. For example, typically, one multifamily unit requires less wastewater treatment capacity, measured on either average annual or peak use, than one single-family residential unit because it serves, on average, fewer people per unit and,

TABLE 10.4 Example of System Development Charge Schedule

Meter size, mm (in.)	Meter capacity, m^3/d (gpm)	Capacity ratio	System development charge
15.9 (5/8)	109 (20)	1.0	$500
25.4 (1.0)	273 (50)	2.5	$1,250
38.1 (1.5)	545 (100)	5.0	$2,500
50.8 (2.0)	872 (160)	8.0	$4,000
76.2 (3.0)	1744 (320)	16.0	$8,000
101.6 (4.0)	2726 (500)	25.0	$12,500
152.4 (6.0)	5451 (1000)	50.0	$25,000
203.2 (8.0)	8722 (1600)	80.0	$40,000

therefore, typically is assigned a lower cost per unit. However, multifamily units could be assigned a higher value based on the class service characteristics of the community. In one instance, a utility located in a ski resort community found that, typically, the peak season number of occupants in multifamily rental units was higher than the single-family residential unit (Corssmit et al., 2002). When peak demands are significant in the determination of infrastructure costs, peaking demands at wastewater collection systems and WRRFs should be considered in the development of fee schedules for various types of developments.

A utility may choose to determine specific capacity requirements to reflect a customer with more extreme uses or potential demands. For example, if a large industrial facility is to locate in the service area, the facility size, capacity requirements, or number of fixtures could be used to establish the fee specifically for that facility. For example, a large airport in a cold climate requiring deicing facilities may impose significant BOD (or COD) loads on the system's treatment capacity. Consequently, the utility may choose to compute the total BOD capacity cost for the new user because the BOD capacity cost is substantially higher than the average system strength. Such exception-type users are recognized in various wastewater utilities. Utilities, in accordance with state laws and regulations, should reserve the option to deal with extreme customers appropriately.

5.2 Geographic Area

Fee schedules may also consider the location of the development in computing fees. Location factors may be based in cases where certain geographic areas exhibit unique costs or service characteristics. In these cases, it is important that the fee structure calculate separate unit costs for the fee areas based on the individual area costs and growth requirements. Geographically differentiated fee schedules may also reflect policy objectives, for example, in the case of downtown revitalization incentives.

6.0 OTHER CONSIDERATIONS FOR SYSTEM DEVELOPMENT CHARGE PROGRAM MANAGEMENT

6.1 Timing of System Development Charge Collection

Implementation of system development charges requires active monitoring and management of the fee program. A utility may either use SDCs to fund capital expenditures or reimburse itself for any growth-related expenditure where working capital or debt is the funding source used as long as the utility's intentions to do so are established before initiating such expenditures. Typically, revenues from SDCs are collected when building permits or certificates of occupancy are issued, which happens as growth takes place. To meet such growth-capacity requirements, however, capital facilities are in place and funds already spent by the time such charges are collected. The timing of collection involves potentially conflicting issues because the utility needs to collect the SDC early enough to make funds available for system improvements. However, the utility can accurately assess SDCs only later in the development process when the actual meter size, use, or number of fixture units is known.

To address timing issues, typically, utilities use bond proceeds to fund large capital projects. Typically, the debt-service payments on these bonds are recovered through wastewater rates and, in some cases, SDCs, given that customer growth and receipts occur as expected. The level of and repayment means, including any legal restrictions, and any rate covenants for outstanding debt need to be considered in the calculation and development of SDCs to avoid potentially double-charging new customers.

6.2 Alternative Fee Calculation

It is recommended that there be a procedure in place for applicants to submit an alternative SDC calculation as well as general procedures for appeals and exceptions. These procedures are required by law in some states. An alternative fee calculation is submitted if the applicant believes that the proposed development will generate effects costing less than the existing fee calculation. Typically, the applicant must demonstrate the claim by providing data certified by a professional in the wastewater industry that proves the development will have less of an effect on the wastewater system.

If the alternative calculation is accepted, the SDC will be reduced to an amount consistent with the alternative calculation. However, if the alternative calculation is accepted, the utility may review actual use of the development once connected to the system to ensure effects to the system are consistent with calculations. If capacity demands to the system are greater than the alternative calculation, the utility may charge the appropriate SDC to account for the increase. Typically, if the alterative calculation is rejected, the applicant may appeal the decision.

6.3 Administrative

Some jurisdictions require that notice of a new or increased SDC be posted within a certain period of time before the effective date of the ordinance or resolution. However, these requirements may not apply if the fee is being eliminated, suspended, or decreased. It is important to ensure that all requirements are met because failure to meet these requirements may result in refunding of collected SDCs. Additionally, refunds may be required if service is not provided in a reasonable amount of time after the fee is paid or if the amounts collected are not used within a reasonable amount of time.

6.4 Accounting and Tracking

Another requirement in monitoring revenues is to ensure that practices are meeting legal requirements. Segregated funds are required by many jurisdictions. To ensure that revenues are applied to intended projects, revenue from SDCs should be placed in a segregated fund earning interest. Fees are to be assessed and collected for the sole purposes identified in the SDC calculations and report. Draws on the fund can be made to pay eligible, growth-related debt service for intended projects. Interest earnings on a specific fund, such as growth-related improvement fees in the combined approach, also need to be applied to growth-related projects. Reliable tracking procedures are essential to ensure revenues from SDCs and interest earnings are used to pay for designated capital projects.

6.5 System Development Charge Review and Adjustment

The fee established for specific capital improvements should be reviewed periodically to determine whether an adjustment is required. Similarly, the CIP and budget should be reviewed periodically to identify growth-related projects. Reporting, reviews, and updates to SDCs ultimately depend on the degree of change in the utility's CIP and state legal requirements.

The utility should also monitor legal activities as they relate to SDCs and continuously work with the public on program administration. System development charges can be effective tools in ensuring that adequate facilities can accommodate growth if they are based on local growth policy, thorough capital planning, established legal standards, and equitable fee calculations and are continuously monitored.

7.0 SUMMARY

System development charges provide wastewater utilities with an additional funding source for growth-related capital projects. They reinforce a "growth pays for growth" policy and give the utility flexibility when developing a financial plan for the future. However, SDCs have a multitude of legal requirements and considerations—from charge design and development to implementation and tracking—that may vary among states. It is necessary to follow and have a strong understanding of these legal requirements for a wastewater utility to have legally defensible and equitable SDCs for use as a capital-funding source.

8.0 REFERENCES

American Water Works Association. (2017). Principles of water rates, fees, and charges (Manual No. 1).

Banberry Development Corporation v. South Jordan City, 361 p.2d 899, 903-03 (1981).

Corssmit, C. W., Malesky, C. F., Mumm, J. G., & Kron, N. F. (2002, September 28–October 2). Wastewater impact fees: A significant legal ruling. *Proceedings of the 75th Annual Water Environment Federation Technical Exhibition and Conference* [CD-ROM], Chicago, IL, United States. Water Environment Federation.

Galardi, D., Buus, B., Cormier, D., & Smith, T. (2004, March 14–17). Development fee trends and Tucson case study. *Proceedings of the 2004 AWWA/WEF Joint Management Conference* [CD-ROM], Phoenix, AZ, United States. Water Environment Federation.

Nicholas, J. C., Nelson, A. C., & Juergensmeyer, J. C. (1991). *A Practitioners' guide to development impact fees*. APA Planners Press.

Porter, D. R. (1997). *Managing growth in America's communities*. Island Press.

9.0 SUGGESTED READINGS

Nelson, A. C. (1995). *System development charges for water, wastewater, and stormwater facilities*. CRC Press.

Ross, D. H., & Thorpe, S. I. (1992). Impact fees: Practical guide for calculation and implementation. *Journal of Urban Planning and Development*, 118(3), 106–118.

11

Planning and Rate Study Implementation

1.0 INTRODUCTION

Implementing a cost-of-service and rate study (referenced throughout this chapter as "rate study") can be challenging, particularly for those without experience or understanding of the process and implementation requirements. Successful implementation depends on proper planning, coordination, and effective communication of the need to adjust rates. This chapter addresses the approach, challenges, and steps to implementation. It emphasizes implementing principal rates charged by a utility for recurring service; however, concepts and considerations also apply to other forms of rate studies (e.g., system development charges or other miscellaneous fees and charges). This chapter begins with a brief overview of the rate-setting authority, summarizes the rate implementation process, and concludes with a detailed discussion of various rate implementation issues.

2.0 BACKGROUND: RATE-SETTING AUTHORITY

The approach and method to setting rates varies widely for utilities. For investor-owned utilities and certain public utilities, typically, the rate-setting process is governed by the requirements and procedures of a public service commission. For most publicly owned utilities, the authority to set and modify rates is granted to locally elected officials or their appointees. Whether the utility is publicly owned or investor-owned, both types of entities must follow the prevailing legal and regulatory requirements to adjust rates. The legal and regulatory requirements may include provisions that regulate, to varying degrees, how rates are determined as well as notification requirements and corresponding opportunities for public comment. As a result, utility organizations implementing a rate study should begin by identifying these minimum requirements.

3.0 EXECUTIVE SUMMARY: RATE IMPLEMENTATION OVERVIEW

The rate study implementation process can be organized into the following three distinct phases: (1) the preliminary or planning phase, (2) the active implementation phase, and (3) the ongoing postimplementation phase.

As Figure 11.1 shows, planning forms the foundation of the rate study implementation process. Planning efforts should focus on the following: determining the underlying goals and rate-making objectives of the rate study, which affect the scope and process of rate implementation; identifying key personnel responsible for implementation (the "project team"); identifying and assessing data required to support the rate study; developing a reasonable timeline with key milestones that coordinate, to the extent possible, with other planning activities (e.g., annual budget); identifying affected stakeholders' interests and concerns; establishing communication protocols, strategy, and methods (the "communications plan"); and identifying a budget for rate-implementation activities, including but not limited to cost

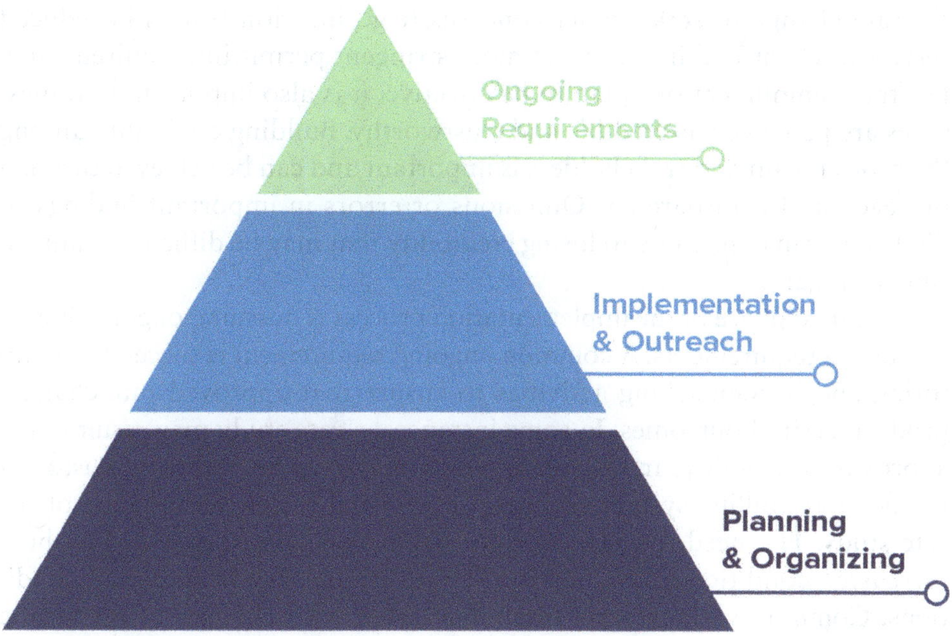

FIGURE 11.1 Implementation Phases

of stakeholder communication and outreach, professional and/or contracted services, and so on (collectively referred to as the *implementation plan*).

The first step of the implementation plan is to identify and engage the project team, identify and compile necessary data, and initiate the rate study timeline. During the performance of the rate study, it is important to ensure that key assumptions, particularly policy assumptions, are well reasoned and supported because they will be an area of focus during rate implementation. After the preliminary findings of the rate study are known, the project team should begin to evaluate the effects on customers, which may result in the review of initial assumptions before developing formal recommendations to mitigate certain effects. In some instances, it may be appropriate to present options to provide decision-makers and the public the opportunity to weigh in on key assumptions.

Once the project team has identified formal recommendations, including alternatives, the project team can begin communicating findings and recommendations based on a communications plan. The communications plan should identify the general approach to communicating findings that provides necessary context and stakeholder engagement, which means options for communicating the findings are varied. Today, the Internet and social media can provide effective ways to engage with stakeholders but can also present challenges. Whatever the means of communication, it is important to link the consequences of implementing or not implementing the rate change to stakeholder concerns (e.g., inaction results in reduced service levels or inability to meet more stringent permitting requirements). For the communications plan to be effective, it is also important that messages are perceived as credible and trustworthy. Building credibility among the project team and stakeholders is important and can be achieved through outreach and transparency. Omissions or errors in important findings or disclosure can contribute to losing credibility that may be difficult or impossible to regain.

The final phase of the implementation process is pursuing ongoing implementation requirements. A common ongoing requirement is successful monitoring and benchmarking activities to ensure that approved rate changes produce desired outcomes. In some instances, a rate study may result in the approval of a multiyear rate plan, which may require recurring customer notifications, billing system changes, or a formal review and update of the rate study. The need to perform a formal review can be assessed by how forecasted conditions, assumed in the rate study, vary from actual conditions. Common variances are often associated with changes to the timing or magnitude of planned expenditures, customer demands, and billed flows. It is important to emphasize that, sometimes, these variances may produce a marginal effect in the current period but may have a material effect over a longer period because of compounding effects. As a result, there may be a need to modify the rate schedule and rate implementation plan to recognize changing conditions.

4.0 PHASE 1: DEVELOPING THE RATE IMPLEMENTATION PLAN

An effective implementation plan includes several key elements, as noted in Figure 11.2. Depending on the complexity of rate implementation requirements, the need can range from a more formal written plan to a less informal verbal understanding. Regardless, the goal of each plan is to aid in identifying and organizing the necessary project team members and activities involved in successful rate implementation.

4.1 Project Team

The project team is responsible for the planning, implementation, and ongoing requirements associated with a rate study. The project team can include individuals both internal and external to an organization; however, the team should have at least one team member internal to the organization who is

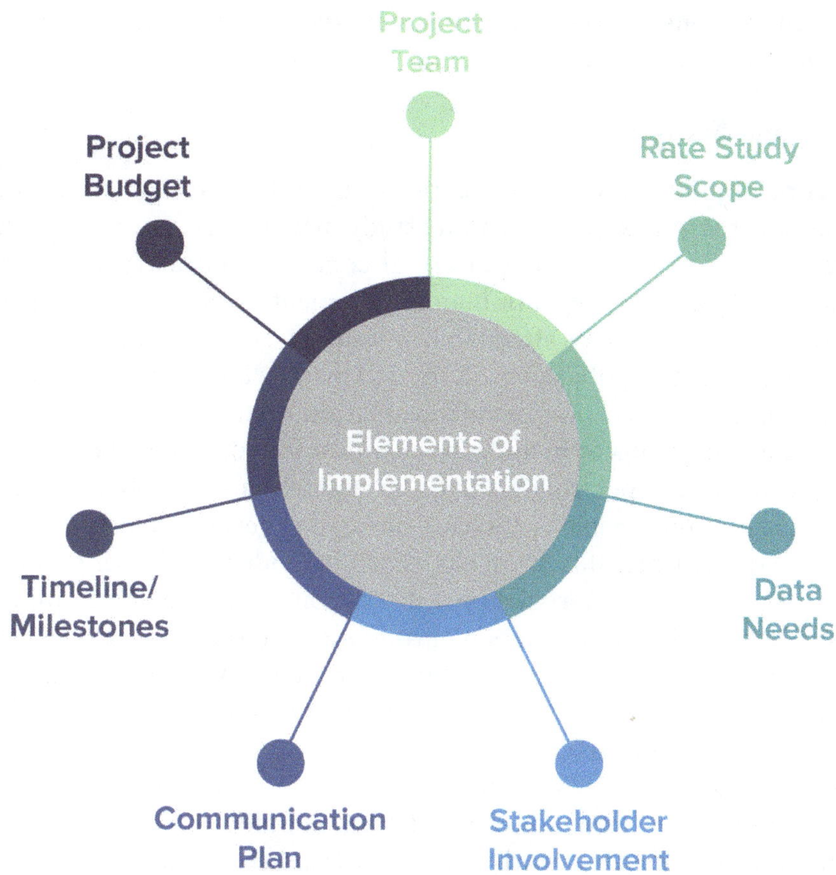

FIGURE 11.2 Elements of the Implementation Plan

responsible for overseeing all tasks (the "project lead"). Ideally, the project lead should have a general understanding of overall utility operations; key staff and their roles, as well as an understanding of the roles of outside engineering, legal, and financial consultants; and timing and requirements of existing planning activities and cycles.

Some utilities may have the necessary expertise and resources to perform the rate study in house; however, because it is a time-consuming activity, it is common for organizations to contract with an outside rate professional or seek other outside technical assistance to perform a rate study. Utilities with limited resources that serve rural or economically disadvantaged communities may be eligible for free technical assistance during the development of a rate study. Interested parties should inquire with their respective state environmental agencies or associated nonprofit organizations to determine availability. During the development of the implementation timeline, it is important to consider the additional time required to procure contracted or technical assistance services.

4.2 Rate Study Scope

The scope of a rate study should consider underlying reasons necessitating it. Common reasons to perform a rate study may include, but are not limited to, deteriorating or improving financial condition; infrastructure funding needs; compliance with legal or loan requirements; or concerns such as rate equity, revenue stability, or affordability. Ranking goals can aid in organizing sometimes competing objectives. Refer to Chapter 8 for a discussion of rate policy objectives and a sample rate evaluation matrix.

Common elements to include in the rate study scope may be a review of historical financial operating results, a forecast of customer billing statistics and demand requirements, a forecast of existing revenues, identification of revenue requirements, allocation of costs to individual customer classes, and design of rates. For a more detailed discussion of revenue and rate design requirements, refer to Chapters 5 through 8. The scope should reflect the complexity of the rate study goals and proposed changes to the rate structure, affecting the effort to assess effects on customers and corresponding communications requirements.

As discussed previously, should a utility contract with a rate professional to perform the rate study, the project team should factor in the time and additional cost associated with issuing a formal solicitation or request for proposal. The identification of rate study goals and scope elements provides the foundation for the solicitation.

4.3 Rate Study Data Needs

The development of the rate study requires significant time to compile data to perform necessary evaluations defined by the study scope. This is particularly true for utilities that do not perform regular periodic updates. A list of required data needs can help to organize requests that must be made to various departments within an organization. Table 11.1 provides an overview of common data needs and sources.

Common challenges with data acquisition include data availability and data quality. The challenges presented by a lack of data availability are obvious, particularly when data are critical to key rate study objectives, such as customer billing data are to rate design. Inaccurate data can present a similar or greater challenge if data are not validated before analyzing or incorporating data to the rate study. Common validation techniques can be to cross-reference data to other data sets or reports. For example, customer billing statistics may include associated charges, which can be cross-referenced

TABLE 11.1 Summary of Common Rate Study Data Needs

Description	Category	Data use	Source
Billing data	Statistical data	Calculation of rate revenues, cost allocations, rate design, and evaluation of customer bill effects	Customer service/billing department (note: may require technical support from billing provider)
Treated wastewater flows by facility or bulk service provider	Statistical data	Determination of variable costs, demands, comparison of billed versus treated flows	Engineering department operating reports
System characteristics (miles of line, number of lift stations, etc.)	Statistical data	Cost allocations	Engineering department
Other revenues	Statistical or financial	Identify other revenues such as assessments, system development charges, pretreatment, miscellaneous fees, grants, etc.	Customer service, finance, or other departments
Fixed-asset records	Statistical or financial	Cost allocation, rate design, desktop asset evaluation, etc.	Accounting/finance departments
Historical financial data (trial balance, audit, cash balances, etc.)	Financial	Evaluation of trends in revenues or costs, comparison of actual to budget, etc.	Accounting/finance departments
Budget	Financial	Identification of current or expected revenues, expenses, and capital outlays	Finance/budget departments
Capital improvement plan	Engineering/financial	Projection of near- or long-term expected capital expenditures	Engineering/budget/finance departments
Loan agreements and debt schedules	Financial/legal	Loan compliance requirements, projection of debt repayment	Finance/budget/legal or other departments
Contractual agreements	Legal	Determining contracted expenses, bulk services, cost sharing, etc.	Administration/legal or other departments
Financial/management policies	Legal/other	Financial compliance evaluation	Administration/finance/budget or other departments
Schedule of rates	Legal/other	Calculation of revenues and customer bills	Administration/legal or other departments
Economic and demographic data	Other	Economic forecasts, new housing starts, etc.	Administrative/planning or other departments
Engineering studies or master plan reports	Other	Identifies long-term capital needs/system characteristics, etc.	Engineering department

against reported revenues. Unexpected variances are a key indicator of a potential issue with data. Consideration should be made to the technical expertise required by the project team member responsible for review and data validation.

4.4 Stakeholder Interests and Concerns

Any plan should specify the stakeholders involved, the nature of their involvement, expected input from stakeholders, whether it is advisory or binding, and the timing of their involvement. Table 11.2 summarizes typical parties whose interests and implementation requirements may need to be anticipated at some point during a rate study.

Successful implementation of a rate study can be affected by how well stakeholder interests are balanced. Certain interests (e.g., regulatory and legal) must be satisfied; however, other interests may be addressed through consensus building for individuals internal and external to the utility. Building consensus is most effective when affected parties are strategically involved in the rate study process at appropriate times and in a manner commensurate with their interests and influence. Achieving consensus can vary in difficulty, and most rate studies occur in complex environments where the chances of success are improved if a well-defined communications and stakeholder engagement plan is developed.

4.5 Communications and Stakeholder Engagement Plan

Like the rate-implementation plan, the communications and stakeholder engagement plan can vary in formality and complexity. The plan should consider several key elements, as noted in Figure 11.3.

4.5.1 Communication Protocols

The process for reviewing and approving information to be disseminated broadly within the utility and to the public should be established. In noncontroversial rate studies involving a minimum number of stakeholders, such protocols will be simple and require little formality. In complex rate studies, clear channels for the review and release of documents are needed. The project lead or other competent senior staff member should be identified as the coordinator for reviewing and approving all communications, including how the distribution of information is handled. In litigious rate studies, legal counsel may have to serve this function whereby attorney–client privilege can be observed.

TABLE 11.2 Common Utility Stakeholder Interests and Concerns

Description	Category	Possible interests and concerns
Administration/ utility staff	Utility	Interested in public/governing body concerns, sustainability, compliance, service to new customers. Can be focused on both near- and long-term issues.
Retail customers	Ratepayer	Quality of service, resource stewardship, affordability, and equity among customer classes. Can be focused on near term.
Wholesale/industrial customers	Ratepayer	Similar concerns to retail customers. Interested in industrial pretreatment requirements/costs. Can be focused on equity of cost allocation in design of rates.
Capacity rights/ contract customer	Regional partner	Interested in service agreement compliance and cost sharing. May also have other vested interest in management, operations, sustainability, capital finance, capacity availability, etc.
Advisory committee	Ratepayer	Similar interests of administration/ utility and ratepayer. Can be focused on both near- and long-term issues.
Elected officials/ regulatory agencies	Rate-setting body	Public perception, rate-making objectives, compliance with regulatory and other legal requirements. Can be focused on near-term issues.
Developers and special interests	Special interests	Have a specific interest directly related to the organization's cause or business interests. Often focused on a specific agenda.
Financial advisors	Utility service provider	Interested in capital finance, debt structure, and financial condition of the utility.
Rate consultants	Utility service provider	Interested in financial sustainability, rate study goals and methods, and rate implementation.
Consulting engineers	Utility service provider	Interested in capital improvement plan, infrastructure requirements, regulatory compliance, and capacity needs and use.
Legal counsel	Utility	Interested in regulatory and legal requirements.

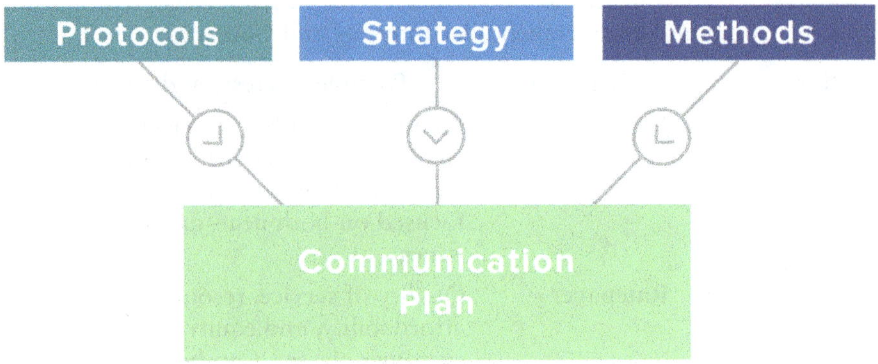

FIGURE 11.3 Communications Plan

4.5.2 Communication Strategy

An effective communications strategy identifies the approach and message to support implementing study recommendations. The following key elements of the communication strategy were adapted from the *Rate Approval Process Communication Strategy and Toolkit* (The Water Research Foundation, 2016), which provides a comprehensive resource and reference document for the development of the communications plan:

- Identify the basis or need to adjust rates and consequences of inaction,
- Link consequences to customer or stakeholder interests and values,
- Tailor communication to the understanding and needs of the stakeholders, and
- Build credibility and trust through outreach and transparency.

Research indicates that active public engagement during rate implementation results in a greater likelihood of successful implementation. Effective engagement is enhanced by creating an emotional connection. The emotional connection is made when the messaging provides strong context to the concerns or interests of stakeholders (i.e., inaction will lead to reduced service levels). Conversely, context-weak messaging (i.e., the utility needs additional revenues) does little to help stakeholders connect to the rate-implementation need. Tailoring communications to the specific needs of stakeholders, such as explaining the utility environment (e.g., the background of legal or regulatory requirements), will help stakeholders to better understand complex issues and study objectives. Credibility and trust are necessary for stakeholders to engage and accept the message related to rate implementation. Communication about rate needs must be consistent, factual, and transparent. Providing access to supporting materials, which can be verified (i.e., disclose assumptions, methodology, and calculations), encourages transparency.

4.5.3 Communication Methods

The communication method should link to the strategy and approach anticipated for rate implementation. At a minimum, the utility will be required to provide public notice of a rate hearing, share or present documentation supporting the need for the rate change, and provide an opportunity for public comment. The use of social media can supplement traditional means of communication; however, the utility should have a formal policy for using such media. *Social Media for Water Utilities* (The Water Research Foundation, 2017) provides guidance and best practices related to social media use. Other means of outreach that aid in communicating the need to adjust rates and address stakeholder interests include several different types of meetings, described as follows:

- Staff meetings. Staff meetings are an integral part of any project. Meeting with staff from other departments is fundamental in communicating the scope of the study, soliciting input, and explaining potential changes at the outset of the study and, later, as results become available.
- Legal briefings. Because the implementation of rates, regardless of how contentious stakeholder involvement may be, always involves legal requirements, the level of involvement by legal counsel should be considered during the study. Legal counsel should be consulted during the planning process to minimize potential challenges. Those responsible for implementing the rate study's results must know the legal requirements for public meeting notices and preparing enabling resolutions or ordinances. If a legal challenge is initiated by ratepayers, legal counsel's advice as to the defensibility of the rate analysis will be invaluable.
- Workshops and work sessions. Presentations at regularly scheduled public meetings of city councils and boards of directors are opportunities to receive guidance from elected officials and input from the public. These presentations can be coordinated with public hearings required at the end of the rate study when the official adoption of rate ordinances or resolutions occurs. However, it is often preferable for workshops and work sessions to be hosted throughout the rate study to create ongoing dialogue, facilitate open communication, and gain consensus for study results.
- Elected official briefings. Typically, meeting individual elected officials can occur without requiring a public hearing. Seek candid input and guidance at these meetings. Briefings can be effective throughout the

rate study process. These briefings can be used not only for discussing key policy issues but also for consensus building around an emerging alternative and educating decision-making parties as to the financial and regulatory rationale for each alternative.

- Public outreach programs. Presentations can be made at meetings of community groups such as the Chamber of Commerce, homeowners associations, and service groups. These meetings should be used to explore community perceptions about the quality and level of service, possible improvements, the value of services provided, and their willingness and ability to pay increased rates to receive the same or improved service. Information from the public is valuable to elected officials, who must balance stakeholder concerns, fiscal requirements, and discretionary expenditures.
- Citizen advisory committee meetings. Committees representing a broad cross-section of the community (affordability advocates, social service providers, business interests, government accountability groups, environmentalists, customer representatives, etc.) can be appointed to participate in the rate study. Periodic meetings are held to present and review study tasks and results as they are developed. The committee should be provided sufficient instruction in rate-making so that members can understand how the results are affected by different assumptions and how compromises may be shaped through trade-offs. Establishing a committee should be considered, especially when there are long-standing concerns that are hindering decision-making. When issues of fairness threaten to become litigious, committees can also provide a forum for dispute resolution.

4.6 Timeline for Implementation

The development of a timeline can be a useful planning tool to coordinate necessary activities of the rate study and implementation schedule. Utilities contemplating a rate study should consider initiating planning during the budget-preparation process preceding the year the rate study implementation is planned. This will provide ample time to develop the implementation plan and budget for any potential expenses associated with implementation. Moreover, the performance of a rate study relies on the availability of timely information for the findings to be relevant. The timeline can aid in identifying the optimal point at which a rate study should be initiated. Figure 11.4 provides a sample timeline related to rate study implementation.

As shown in Figure 11.4, rate-implementation activities are scheduled around important utility events that produce necessary data (e.g., budget documents, capital budgets, engineering studies, and financial reporting).

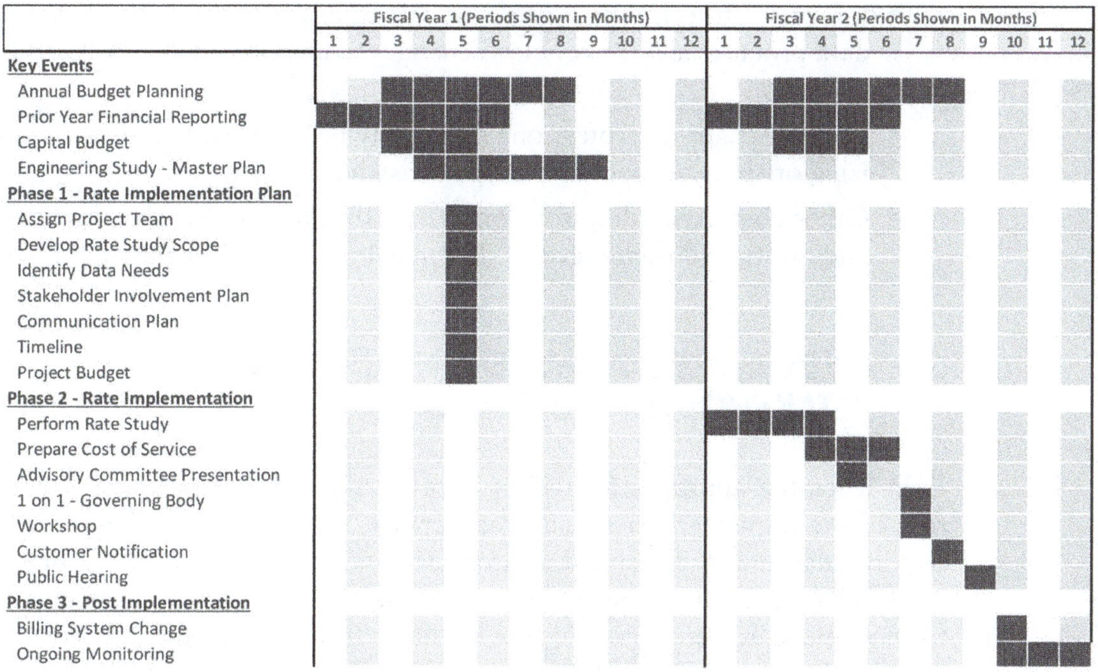

FIGURE 11.4 Sample Timeline

Note that actual scheduling will rarely follow initial plans because of emerging issues and unforeseen delays, which are almost inevitable. The time required for internal review and external interaction with stakeholders can also be difficult to predict, even with careful planning. Including adequate contingencies in planning can aid in successful implementation.

4.7 Project Budget

The cost to perform a rate study can vary depending on the circumstances, scope of work, and if the utility intends to use outside professional or contractual services. The project budget should consider cost estimates for the following activities that may require outside consulting or professional services:

- Preparation of the rate study by a rate professional based on the rate study scope;
- Preparation, review, or prioritization of capital needs and associated project cost estimates by a consulting engineer;
- Legal review of the rate study and resultant tariff by an attorney to assess compliance with applicable regulations or laws;

- Information technology professionals to aid in updates to websites, data procurement, or revisions/changes to customer billing/reporting software;
- Communications professionals that may be required to aid in organizing or sponsoring community outreach events; and/or
- Other costs, including for communication materials, advertisement and notifications, and events that may be required for outreach.

5.0 PHASE 2: PLAN IMPLEMENTATION AND STAKEHOLDER OUTREACH

Even when the appropriate level of effort is expended in advance, execution of the implementation plan may present unforeseen challenges that require adaptive management strategies. Rate studies are prone to unexpected changes because the analytical process reveals information that calls for midcourse adjustments. Because of this, a flexible implementation style is desirable. The following discussion provides an overview of the analytical steps of the rate study, the consideration and management of alternatives, and other challenges and additional implementation requirements.

5.1 Rate Analysis

A rate analysis may have as many as three distinct analytical steps. These steps—revenue and revenue requirement projections, cost-of-service allocations, and rate structure design—recognize key assumptions that, as revised to match differing goals or objectives, can produce many alternatives and outcomes. A solution to this challenge is to rank alternatives through a two-step process where each alternative is assigned a weighted ranking based on the product of how the alternative ranks in achieving a particular objective and how each objective is ranked in terms of overall importance or need. The ensuing discussion provides an overview of the principal analytical activities of the rate study and key assumptions that are likely to produce alternatives. Ultimately, the goal is to limit the alternatives to those that present a meaningful comparison among policy options and prioritization of rate-making objectives.

5.1.1 Revenue and Revenue Requirements Projections

In forecasting utility revenues and revenue requirements as outlined in Chapter 5, the project team must develop a forecast of customer billing statistics (i.e., accounts by customer class, billed flows, and treated wastewater flows) that affect revenue projections and identify assumed escalation factors or variables affecting the projection of operating expenses and capital needs and financing options, recognition of debt service payments, and any required deposits to fund reserves according to policy or debt-covenant requirements. Additionally, for combined water and wastewater utilities, the project team will need to allocate such costs among water and wastewater functions.

Key assumptions that relate to policy choice and may produce different alternatives are often attributable to the assumed mixture of debt versus cash financing, priority and timing of capital projects, extension or expansion of service, and how such costs are funded (i.e., how such costs should be balanced among existing and new customers). As detailed in Chapter 3, developing a formal policy that is approved by the governing body of a utility related to financial-management results and utility rate-setting goals (e.g., the maximum ratio of debt to net facility, minimum cash reserves, and debt service coverage requirements) can help minimize challenges to key assumptions and present an opportunity for stakeholder engagement.

Changes to other assumptions (e.g., cost escalation factors, cost of capital, and timing of capital improvements) can also result in alternatives but, typically, are presented to demonstrate sensitivity and gauge stakeholder risk aversion. This is an important consideration because decision-makers may challenge implementation if they perceive that rate recommendations are based on flawed assumptions that may be overly conservative (e.g., cost escalation assumptions are too aggressive relative to historical trends) or not sufficiently conservative (e.g., growth assumptions are too high). An effective method to address these concerns while limiting potential alternatives is to assess the sensitivity of the upper and lower limits of likely outcomes sometimes referred to as "book-end assumptions." Alternatives for presentation may be limited to only those that produce meaningful differences when comparing book-end assumptions.

5.1.2 Cost-of-Service Allocations

Assignment of revenue responsibilities largely depends on appropriate cost-allocation methods, applicable waste stream strength parameters, joint versus specific cost assignments, and direct cost assignments, as outlined in greater detail within Chapters 6 and 7. The specific approach to cost-of-service allocations is often technical and data-driven, so challenges may be limited to either overarching assumptions about appropriate rate-making methods (e.g., commodity demand method versus base extra capacity method) or assumptions that may affect specific stakeholders such as legal counsel's interest in compliance of regulatory requirements or wholesale customer interest in overhead or other cost allocations affecting user charges.

5.1.3 Rate Structure Design

Several potential rate structures (e.g., those involving the use of fixed versus variable charges and quantity allowances) and rate-making goals may be selected at the study's outset. Often, a principal concern at this stage is the resulting effect on customers. Although there may be consensus on underlying assumptions, the relative effect on customer bills may outweigh previous consensus. A common approach to addressing this challenge is to present an option that phases in rate structure change or achieves study goals gradually over time.

5.2 Stakeholder Outreach

Once a recommendation or set of alternatives emerges from the rate analysis, findings should be documented and presented. Communications and outreach should follow the approach and strategy identified in the communications plan (see Section 4.5 of this chapter). The context to support rate changes can now be identified to form the foundation of the message. Sometimes, communication requirements are not related to a singular issue and may be complicated by proposed alternatives.

The concept of message mapping, which is widely used in risk communication, can be a useful tool for adaptation to rate communication. A message map aims to educate and inform stakeholders to build an informed dialogue about important decisions. A message map contains three tiers of information, including (1) the stakeholder to whom the message is addressed, (2) the anticipated stakeholder question or concern, and (3) up to three key messages in response, with up to another three supporting statements or pieces of information (Lin & Petersen, 2007).

5.3 Rate Adoption

As previously discussed in the overview of rate-setting, typically, the legal process for rate implementation consists of public notice requirements and legal procedures that vary from state to state. The existing legislative instrument can often be updated without extensive revision. However, it is often the case that existing legislation contains outmoded, undocumented, vague, or inaccurate sections. There should be no hesitation in making whatever revisions are required to remedy historic deficiencies. Presumably, careful consideration was given early in the rate study to ensure that adopted rates can be readily implemented and administered continuously.

5.4 Billing System Modifications

Typically, rate studies result in changes that must be reflected in customers' bills. The billing system may need to be revised to accommodate the following requirements:

- Changes to the rate structure (e.g., addition of a new fixed charge or differing charges for a new class of customer);
- Reclassification of customers, particularly commercial customers, for any changes in the nature of the customer's business activity that may alter waste-strength-loading assignments or other bases for customer classification; and
- Implementing flow-based charges, which could require using average winter water use for residential customers and ongoing water use for nonresidential customers.

In some cases, the wastewater utility may not handle billing directly. For example, the charge for sewer service may be included on the tax rolls as part of annual property taxation. In this case, the agency must create a customer database that correlates water supplier account numbers with taxing authority parcel numbers. This billing process requires a high level of coordination among the wastewater utility, water utility, and taxing authority.

In some cases, billing systems may have limited capabilities, which should be determined at the beginning of the study so that a rate structure that the billing system cannot implement is not selected for further study. From time to time, new billing systems are installed in conjunction with the rate study. In this case, extraordinary care must be taken to limit disruption to the utility's day-to-day billing processes, enable the collection of customer billing data that are required for the rate analysis, and ensure successful implementation of new billing systems and rates.

5.5 Customer Service Requirements

Customer service staff can include accounts receivable and utility billing staff as well as customer service representatives who have direct contact with customers over the telephone or in person to discuss billing and payment matters. In addition to these staff, field staff responsible for meter reading can play an important role in utilities that also provide water service. For example, meter readers should be instructed to advise the billing staff of significant changes in commercial business types, which might require a reclassification of customers from one waste strength category to another. To ensure that customer service is thorough and accurate, training is required in appropriate areas, as follows:

- Accounting and utility billing staff need to understand the system of rates and charges to bill customers correctly. Training in the use of billing system software is a prerequisite.
- Customer service representatives need to understand the rate structure and billing system so that billing questions can be resolved. Knowledge of the billing system is a prerequisite. Knowledge of the provisions for handling exceptions in relevant policies is also essential.
- Meter readers need to be familiar with the rate structure to notify billing staff of any significant changes in commercial business that would affect billing.

Training efforts should occur at the end of a rate study and periodically thereafter for new staff. It is advisable to have customer service staff attend public presentations of the rate study to familiarize them with the same information that ratepayers and elected officials receive. This training can also be interactive. In other words, the experience of accounting and billing staff that could be influential in subsequent rate studies should be compiled and presented in the next study. For example, implementing a new rate structure may enhance billing processes. Simple modifications in the rate structure (e.g., rounding rates) may be all that is required to lighten demands on billing system staff. It is advisable at the kick-off meeting to a rate study to have customer billing staff convey their experience with the billing system and customers' perceptions.

6.0 PHASE 3: ONGOING IMPLEMENTATION

Ongoing implementation requirements can be associated with monitoring and benchmarking results to ensure that objectives of the rate study are met. In some instances, approval of a multiyear rate plan may require recurring customer notifications, billing system changes, or a formal update or review of the initial rate study. This section provides an overview of these ongoing implementation requirements.

6.1 Monitoring and Compliance

An important goal of a rate study is that approved rates support the utility's financial sustainability, maintain compliance with financial policies and covenants, and achieve rate-making objectives. Measurement and monitoring are necessary to assess compliance with these objectives. For example, managers need to determine that they receive the correct revenue from implemented rates. Revenue from ratepayers may differ from what is projected in the rate analysis if there are discrepancies between data used in the rate analysis and actual customer billing data. Summary revenue reports from all sources can be compared with rate models to identify adjustments needed to calibrate rate models or rectify billing system issues.

In some instances, as an outcome of the rate study, a utility may identify data that need to be tracked going forward. For example, data on water resource recovery facility loading, industrial customer sampling, and other sampling may be necessary to validate cost-allocation assumptions made without actual data, which will require validation.

Tracking other financial or operational data is recommended to assess the utility's financial and operational sustainability. *Effective Utility Management: A Primer for Water Sector Utilities* (U.S. Environmental Protection Agency et al., 2024) provides suggested frameworks for monitoring and benchmarking.

6.2 Future Updates and Other Considerations

Assumptions relied on in forecasting customer demands, cost inflation, timing, and the level of capital spending and design of rates, among other assumptions, may vary with actual conditions. The extent to which projected results vary from results is a key indicator that there may be a need to perform an update to the rate study. It is important to note that some variances can compound over time, and careful consideration should be given to variances that may be recurring in nature (e.g., reduced customer demands and billed flows).

Sometimes, a utility may have a phased multiyear rate plan that requires recurring customer notifications, billing system changes, or a formal rate review. Other considerations for future rate study updates may include updating the initial study for data that may have been lacking or delayed in availability. For example, referencing Figure 11.4, if the master plan was developed after the rate study, updating revenue requirements to reflect any changes or updates to the capital improvement program would be a prudent consideration.

7.0 CONCLUSION

The rate study and resulting proposed rate structures are undertaken to determine how much each class of customers should pay. Typically, wastewater utilities adopt rates through a public hearing process that allows various stakeholders to have a voice in establishing proposed rates. Each utility will have its own unique circumstances that should be evaluated before starting a rate study so that ample opportunities are provided for internal and external stakeholder input, review, and comment. This will enable the decision-making body to gain stakeholder support and render well-informed decisions.

Developing a rate study implementation plan, creating an effective communications strategy, and establishing criteria to monitor study results on an ongoing basis will aid in successfully communicating and implementing desired changes to utility rates.

8.0 REFERENCES

Lin, I., & Petersen, D. D. (2007). *Risk communication in action: The tools of message mapping*. U.S. Environmental Protection Agency.

The Water Research Foundation. (2016). *Rate approval process communication strategy and toolkit* (Web Report No. 4455a).

The Water Research Foundation. (2017). *Social media for water utilities* (Report No. 4638).

U.S. Environmental Protection Agency, American Water Works Association, Association of Clean Water Administrators, Association of Metropolitan Water Agencies, Association of State Drinking Water Administrators, National Association of Clean Water Agencies, Water Environment Federation, & The Water Research Foundation. (2024). *Effective utility management: A primer for water sector utilities*. https://watereum.org

12

Affordability

1.0 INTRODUCTION

This Manual of Practice (MOP) is focused on wastewater utility pricing. Wastewater charges, however, are often billed alongside water and/or other utilities. This chapter's discussion, similarly, focuses on billing and affordability for wastewater service, but readers should consider the broader suite of utility services billed together when interpreting affordability metrics and thresholds.

Customer affordability, or simply, affordability, is a customer's ability to pay for a service. As wastewater utilities seek to develop rates and charges in line with their basic pricing objectives, another objective—affordability of wastewater service—has become important for elected officials, utility managers, and the public. As system policymakers plan for the future, a significant consideration is whether proposed rates and charges are within or beyond the financial capacity of ratepayers. Though the financial capacity of ratepayers will vary among utility service areas, and the definition of "affordable" has multiple perspectives, this chapter will provide information about how to begin the affordability discussion at the utility. It will address various considerations for what affordability means and provide opportunities to enhance affordability as well as provide references to explore the topic further.

Within this chapter, we will discuss affordability, or maintaining affordable rates, as a utility concern, recognizing, of course, that it begins as a customer (individual, household, or business) concern that ultimately affects the utility when customers cannot, or do not, pay utility bills. Rates and charges that are not affordable for a segment of ratepayers can lead to opposition to utility rate adjustments, increasing accounts receivable, rising expenses for uncollectable accounts, reduced level of service, and general customer dissatisfaction with overall utility service. As public service providers, it is sometimes also perceived as a utility's responsibility to avoid placing an undue financial burden on its ratepayers. This, too, makes affordability a utility concern.

Most utilities are likely to face an affordability challenge to some degree. That is, at least a small number of customers may struggle to pay for service. The challenge is exacerbated when utilities must raise rates more than economy-wide inflation/growth to meet regulatory requirements, provide for growth needs, or replace aging infrastructure. Challenges associated with maintaining affordable rates also increase when a local economy loses strength, reducing the ability of customers to pay utility bills, even if the rates are held steady.

As the need for investments in utilities continue and grow, so too is the importance of an open conversation on ways the investments are paid for and who ultimately pays for them. In early 2024, the U.S. Environmental Protection Agency (U.S. EPA) published its 2022 Clean Watersheds Needs Survey (https://www.epa.gov/cwns/clean-watersheds-needs-survey-cwns-2022-report-and-data), which identified $630 billion in wastewater and stormwater system capital and investment needs. This is the estimated cost to maintain and modernize publicly owned wastewater collection systems and treatment works, stormwater infrastructure, nonpoint source control, and decentralized wastewater treatment systems. Although these high-level costs are important and impactful, costs will be borne primarily by customers of local utilities and make affordability a significant issue that utilities must consider and deal with as they undertake capital and operating programs.

Affordability is a key criterion for virtually all stakeholders when utilities seek successful implementation of rates and charges. Ratepayers need to be able to pay their utility bills without adversely affecting their ability to meet other important financial needs. Typically, decision-makers find that discussions on affordability improve the public's acceptance or approval of proposals to increase rates. Incorporating these considerations to the utility's long-term financial planning can show that they are making prudent choices in the best interest of the utility and its customers. As rates are developed under key cost-of-service-based principles, it is important to examine the community's ability to pay proposed rates because of the diversity of interests that are potentially affected. This chapter provides stakeholders the information needed to have a meaningful conversation regarding affordability.

2.0 AFFORDABILITY STANDARDS

There is no single standard or definition of what constitutes affordability for wastewater utility rates and charges. For the purposes of this manual, affordability is described as the ability of the utility's customers to pay their utility bills without undue hardship. In this context, undue hardship is the inability of customers to pay for other basic needs as a result of paying the utility bill. For example, affordability for a residential customer would mean the ability to pay the utility bill and to pay for food, housing, normal medical needs, and other basic necessities. For commercial customers, affordability is the ability of the customer to pay the utility bill in addition to other basic or essential business expenses. In practice, this ability to pay is then often expressed as a specific, numeric threshold beyond which rates and charges are considered unaffordable for vulnerable customers.

Given the large spread in demographic characteristics across a community and utility service area, it is unlikely that an entire service area would fail a standardized affordability test. However, it is likely that every utility will have unemployed or low-income customers who would find almost any bill difficult to pay. Similarly, families with extraordinary medical expenses might be unable to pay for even modest utility charges. As a result, providing utility managers and decision-makers with the information they need to understand the overall makeup of their customer base and the potential range of ability to pay within their community is beneficial as they formulate rates and charges proposals and consider financial-assistance programs to mitigate the number of these situations and further enhance affordability.

Part of the issue of affordability is that, typically, wastewater bills are seen as competing with other basic utility expenses, such as gas or electricity, for limited customer dollars. To the extent that other utility bills increase, the available customer dollars to support wastewater increases are reduced, and vice versa, without growing incomes. Because affordability is often viewed in light of total utility expenses, as water, electricity, natural gas, and telephone bills increase rapidly for various reasons it becomes more difficult to raise wastewater rates and have them remain affordable. Although wastewater rates should be set to recover the needed expenses of the utilities based on cost-of-service principles, this competition does exist and provides needed context within the discussion of affordability.

As a result, noting what makes your community unique and individual challenges that face your utility customers are critical to evaluating a standard for affordability for your utility. This information together with published research and thresholds (or identified levels of selected ratios) on affordability discussed herein will aid your utility in identifying the best approach to fit your circumstances.

3.0 AFFORDABILITY THRESHOLDS

Because of the subjective nature of affordability, the differing distribution of financial resources among customers, and the wide variety of circumstances that significantly affect affordability of utility services, there will likely never be a universally accepted threshold at which these services become unaffordable. Changes in taxes, fees, and the prices of other goods and services may also affect the ability of utility customers to afford wastewater rates and bills. In other words, considerations of affordability must take into account a dynamic set of factors affecting the claims on customer resources and your utility's specific circumstances. Nonetheless, there are sources of guidance that can be considered when evaluating affordability, defining affordability, and formulating assistance programs.

3.1 Charges Relative to Median Household Income

The most frequently cited affordability guidelines are those from U.S. EPA. This federal agency, which mandates certain standards for water and wastewater quality and provides loan and grant programs, has recognized the need to consider expensive infrastructure programs in the context of a community's ability to pay for those improvements. This system affordability concept differs from customer affordability in that it does not contemplate the ability of individual customers to pay their bills, only the financial capacity of the collective customer base to pay for capital upgrades. For example, U.S. EPA mandates certain aspects of wastewater user charges for Construction Grants Program recipients and those charges have stayed relatively consistent over the last few decades. In these user charge guidelines, U.S. EPA allows grantees to establish low-income user charges for residential customers, where *low income* is defined as any residence with a household income below the federal poverty level (Grants for Construction of Treatment Works, 2003).

With respect to a system affordability threshold beyond which utility bills might be deemed unaffordable, U.S. EPA has offered guidance (first in 1997 and upheld more recently) that affordability of investment in wastewater systems can be defined as 2% of median household income (MHI) in addition to other factors to be examined in assessing system affordability (U.S. EPA, 2021). These factors include such items as the debt load of the system, local socioeconomic conditions, and other relevant financial considerations and are included collectively in a financial-capability assessment. This threshold of 2% of MHI is applied on a systemwide basis and is not intended to determine customer affordability of user charges on an individual household or customer basis. For reference, the U.S. Census Bureau reported that median household income in the United States was $80,610 in 2023 (U.S. Census Bureau, 2024).

Before issuing wastewater guidance, U.S. EPA drafted system affordability criteria for meeting water quality standards that are still in place today (U.S. EPA, 1995). This interim guidance suggested that beyond 2% of median household income, the pollution control cost "may place an unreasonable burden on many of the households within the community" (U.S. EPA, 1995). These criteria are for evaluating the cost of implementing potential water quality programs rather than the possible burden of the overall utility charges. U.S. EPA goes on to add that various secondary criteria should be examined to fully assess the financial capability of the system. These are shown in Table 12.1.

Even within U.S. EPA, these levels act as guidelines. As a result, the Affordability Review Panel of the U.S. EPA Science Advisory Board suggested

TABLE 12.1 U.S. EPA Community Financial Strength Indicators

Financial indicator	Strong (3)	Midrange (2)	Weak (1)
Bond rating	AAA-A (S&P; Fitch) or Aaa-A (Moody's)	BBB (S&P; Fitch) or Baa (Moody's)	BB-D (S&P; Fitch) or Ba-C (Moody's)
Net debt/property market	<2%	2%–5%	>5%
Unemployment rate	>1% below the national average	±1% of the national average	>1% above the national average
Median household income/adjusted national MHI	>25% above the adjusted national MHI	±25% of the adjusted national MHI	>25% below the adjusted national MHI
Property tax revenues/total property market value	<2%	2%–4%	>4%
Property tax collection rate	>98%	94%–98%	<94%

that system affordability determinations be made on a regional or local basis rather than nationally (U.S. EPA Science Advisory Board, 2002). Individual states have used this guidance as a baseline to either adopt or inform separate thresholds. Massachusetts Department of Environmental Protection (1997) and Ohio Public Works Commission (1999) are examples of state-level agencies have adopted similar standards for system affordability.

As mentioned in U.S. EPA's guidance, other critical financial tools for utilities take into account affordability; for example, rating agencies take into account affordability of the systems' rates and charges when providing bond ratings. This works as a two-way street for utilities to evaluate their affordability. Fitch Ratings, Inc. (2024) discusses how they measure affordability in "U.S. Water and Sewer Revenue Bond Rating Criteria": "Fitch calculates an affordability rate to determine the number of people whose bill accounts for an outsized portion of their income. Generally, a combined water-related bill that is greater than 5% of household income (or individually, 2.0% for water, 2.5% for sewer and 0.5% for stormwater) is considered unaffordable."

A recent nationally representative wastewater rate survey provides data on wastewater bills as of 2023 (American Water Works Association & Raftelis Financial Consultants, 2024). It indicates that, in the United States, the average residential wastewater charge for the systems they sampled was \$60.78 per month for a residential customer using 1000 cu ft (7480 gal) (note that 1 Ccf = 2.83 m^3). For a residential customer using 1500 cu ft (11 220 gal), the average charge was \$82.09. For a customer using 3000 cu ft (22 440 gal), the average charge was \$142.10. Median charges for these three respective levels of use were \$56.26, \$75.13, and \$131.95, respectively. These results are primarily lower than the recommended affordability threshold figures, suggesting that wastewater charges on a national basis were affordable.

This conclusion should not be extrapolated to conclude that charges are always affordable. Specific local circumstances, including income levels, could indicate ability-to-pay concerns or issues. Also, affordability of wastewater charges may change when conditions change. This could be true particularly during periods of increasing investment in new and replacement infrastructure.

3.2 Other Thresholds

Though the ratio of wastewater costs to median household income is the most common way to evaluate affordability across the full customer base, affordability analysts offer other metrics for consideration as well. Two additional approaches are offered by Teodoro (2024), including the use of an affordability ratio, which calculates utility costs as a share of disposable income rather than total income. Disposable income is calculated using U.S. Bureau of Labor Statistics Consumer Expenditure Survey data (https://www.bls.gov/cex/) to subtract other essential living costs (housing, food, healthcare, energy, taxes). Teodoro (2024) also measures utility charges relative to the required number of labor hours at the local minimum wage to pay those costs, which puts the affordability question in another context.

Two large utilities, Baltimore, Maryland, and Philadelphia, Pennsylvania, consider individual household income with respect to the federal poverty level, creating an individualized assistance calculation so that individual customers receive affordable utility bills.

The U.S. Low Income Household Water Assistance Program (LIHWAP) was the first-ever federal emergency program to provide assistance to families struggling to pay their water and wastewater bills. The LIHWAP was authorized and funded by the Consolidated Appropriations Act of 2021 and the American Rescue Plan Act of 2021 during the COVID-19 pandemic. The LIHWAP was authorized as a temporary emergency program to meet the needs of households with low incomes and during the pandemic. The LIHWAP grant recipients were required to use 150% of the federal poverty level, 60% state median income, or another poverty threshold for determining a household's income eligibility. The LIHWAP funding closed out in 2024, with a total of approximately $1.12 billion awarded and 1.7 million households assisted. As of 2024, efforts are underway to introduce a bill to create a permanent, national water assistance program under the LIHWAP Establishment Act.

3.3 Affordability for Hard-to-Reach Customers

Affordability thresholds are even more difficult to define for households that are not direct customers of a utility. As such, less attention has been given in the literature to the ability of multifamily residential customers to pay for wastewater or other utility services. These customers have a broad range of available financial resources, range of rent/utility payment structures, and other factors related to "ability to pay." Thus, it is challenging to define affordability for these customers and fewer definitions exist. However, there is every reason to expect that high utility bills can have as serious an effect on these customers as on other residential customers.

The Water Research Foundation (WRF) addresses multifamily customers and other hard-to reach customers in *Customer Assistance Programs for Multi-Family Residential and Other Hard to Reach Customers* (2017). This study focuses on indirect customers (those who receive water service but do not directly receive a bill). Often these are renters in multi or single-family residences who pay for utilities as part of rent. Renters present an additional challenge for utilities wishing to define and address affordability issues because there is no direct relationship between the utility and the potential low-income customer, or recipient of the service. According to this report (TWRF, 2017), these households are more likely have lower incomes, not speak English, and move more frequently. Additionally, housing costs can be a higher burden as a percent of income for these customers, meaning even less income is available for other essential costs than for other customers. Utilities with known nonresidential affordability concerns may need to consider creative solutions, some of which are described below.

4.0 OPPORTUNITIES TO ENHANCE AFFORDABILITY

Despite the challenges in defining affordability or affordability thresholds, there are an array of methods to improve the affordability of wastewater services for utility managers who choose to do so. From modifying rate structures to implementing low-income assistance programs, these options have differing effects on customers and, on the utility and, therefore, any affordability program should be tailored to meet specifically defined community objectives. As a result, many utilities use multiple methods to address affordability issues within different populations and from different directions.

4.1 Rate Structure

The rate structure itself may be selected with affordability considerations in mind. If the current rate structure is a flat rate structure (i.e., the bill is fixed and does not vary with level-of-service use), then moving to a (water-use) volume-based approach could assist with affordability. The reason is that if the structure is volume-based, a customer is better able to control the bill by altering the level of use. If affordability is a concern, then reducing the water use will lower the bill. Where wastewater systems already have both a volume-based rate structure and fixed charge-per-billing period, components of a fixed charge could be shifted to a volume charge. By moving more of the cost recovery into the volume rate, the customer has more opportunity to influence a bill by reducing the level of use. This approach is limited, however, because minimum sanitary needs must be met, and the need for revenue stability may limit the extent to which a utility can move cost recovery to a volumetric charge.

An increasing block rate structure for wastewater services can also have the effect of improving affordability for customers most in need of assistance. Such customers are often small users, such as residential and small businesses, and an increasing block rate structure results in the lowest block rates being applied to their use. However, this approach may exacerbate affordability issues for larger nonresidential customers and residents with large or extended families living in the same household. A related option is to charge increasing block rates for water service. This water rate structure would lower water bills for many of the same customers and enable a customer's resources to go farther paying for overall utility services. Furthermore, to the extent that water and wastewater services are on the same bill, the affordability of the combined bill can more easily be targeted and addressed by local officials through a combined rate viewpoint. Of course, where water and sewer providers are different, this may be more complicated as they have to share data readily.

One important reminder is that small users are not always low-income users. Conversely, large users are not always high-income users. For example, low-income users may have poorly maintained or older high-flow fixtures that result in higher use. Therefore, often, it is helpful to evaluate the affordability effects of potential rate structures by specifically considering information on a utility's low-income customers and their use patterns.

Some utilities use a residential wastewater rate structure that caps the level of water use that is subject to the wastewater charge. The intent of this cap is to approximate the maximum water use that enters the wastewater system from these customers. A volume cap can also help enhance affordability because it limits the wastewater bill, and customers are better able to financially plan for their bills if they understand what the maximum could be. However, a cap on use charges may reduce the utility's revenues and can necessitate an increase in rates to recover the same revenues. Additionally, if low-income customers are already smaller volume users, a cap may not provide the desired relief for low-income customers.

A "lifeline" rate structure is one that charges a low rate for a minimum level of utility service that is deemed to be essential for public health needs. Often, a lifeline rate is below the average for the residential cost of service because not all of the peaking cost is reflected in the rate. In such cases, a subsidy comes from a small increase in the rates for other customers of the utility. By specifically providing a rate that is affordable to all but the lowest income customers, the overall affordability of utility service is maximized for those who can least afford it. Sometimes, eligibility through proof of qualifying low income is required to receive this preferential rate. This may include income verification or simple documentation that the customer is eligible for other income-based programs like those offered by electric and natural gas utilities.

Because of concerns about administrative mechanisms needed to confirm income eligibility, some utilities choose to charge all customers a low rate for their initial use. This has a drawback in that affluent customers benefit from some of their use being at a rate intended for low-income customers. In other cases, utilities sometimes equate affordability and age, offering discounts to those customers over a certain age. Although there are certainly elderly customers on fixed incomes that may indeed have affordability concerns, age is not a particularly good indicator of wealth or ability to pay. In any case, the objective of a lifeline rate program is to meet basic utility service needs at a price that virtually anyone can afford.

In assisting nonresidential customers, a rate structure that contains a "water audit" rate (a discounted rate offered as incentive for customers to participate in a water audit) can be helpful. Under this approach, the customer can consent to a water audit by the utility or its representative. The audit objective is to identify opportunities for conservation through upgrades to fixtures or appliances or improved practices and provides those recommendations to the customer. If the customer implements the recommendations of the audit, they get a discounted rate. In other words, if the customer has taken all of the recommended steps to limit water use (and by inference, wastewater use as well), then the customer has minimized the use of utility services to the extent possible. This construct aids in affordability as use is lower and that lower use is billed at a lower rate. The main drawback to this approach is that a water audit rate program can be costly for the utility in terms of staff time (in house or contracted), supplies, training, insurance, and other expenses associated with conducting the audit at a customer location and tracking follow-up.

4.2 Targeted Low-Income Program

Many utilities that have attempted to address affordability considerations have used a targeted approach, particularly with respect to low-income residential customers. These programs can include a variety of tools to address this issue. For example, a program could be composed of one or more of the following elements:

- Bill discount. Wastewater bills could be discounted by a given flat amount or by a fixed percentage for eligible households.
- Crisis vouchers. Periodically, such as once a year, a low-income household may receive a voucher to use as a credit on the outstanding or future utility bill. The voucher would be for a fixed dollar value. Eligibility might be limited to an income criterion but could also include demonstrated extraordinary circumstances, such as medical expenses.
- Bill forgiveness and goodwill adjustments. For reasons of customer goodwill or other objectives, the wastewater utility might have a

policy allowing write-offs of a portion of an outstanding utility bill. Such write-offs might come with certain restrictions as to eligibility for bill adjustments and the frequency with which this might be allowable for a customer.

- Targeted conservation. Water conservation programs may be targeted to the low-income population within the service area or small users of all types. The conservation program could include education, distribution or installation of water-saving devices, and other tools to help customers use utility services wisely. By conserving water, customers will be better able to afford wastewater bills if bills are structured based on water use. These efforts help "shrink the problem" by reducing the overall wastewater bill as well as the associated water bill.
- Leak repairs. A leak repair program aids in wastewater affordability if wastewater bills are structured based on water use. As with conservation, it helps to shrink the problem and provides water affordability benefits as well. Some agencies have grant programs for individual home repairs.
- Payment plans. Many wastewater and water utilities offer payment plans for customers whose utility bill balances are beyond their means to pay in full. By providing such payment plans, the utility provides an opportunity for customers to gain better control over their financial situation and increase the chances that the bill can be paid over a longer period of time.
- Social assistance programs. Some utilities encourage customers to contribute voluntary funds for utility bill assistance to needy customers who are unable to pay their wastewater bills. Often, these funds are administered through existing municipal or charitable social assistance programs that make the determination as to the need of the applicant.

By directly targeting an assistance program to low-income customers or financially challenged businesses, the wastewater utility may cost-effectively improve affordability of its rates and charges for those that are least able to afford them.

4.3 Program Administration

For utilities considering implementing a dedicated affordability program to qualifying customers, there are several administrative components that will require evaluation and planning in advance to ensure a smooth rollout.

4.3.1 Eligibility Validation

To meet program goals and provide financial assistance where it is most needed, an income-based affordability program will need to validate financial-eligibility criteria for individual customers. This can include an application process and review of customer income information and other sensitive documents. While some utilities administer these programs internally, many use third parties to administer and verify eligibility. As noted in Section 3.0, there are multiple methods to define affordability, hence there are multiple ways to set eligibility criteria. To establish a consistent and straightforward approach for the customer and maximize efficiency in program administration, water and wastewater utilities often use the same income guidelines as energy-assistance programs, which can be better established (although utilities are also free to choose alternative criteria if those better suit their needs and goals). Utilities can even accept a customer's qualification for another program as qualification for its own program to reduce the customer's burden for applying and its own burden for administration if another program's eligibility requirements match the utility's requirements.

4.3.2 Billing System Capabilities

When considering the appropriate type of affordability program (and the specific benefits offered), the capabilities of the billing system should be evaluated to make sure they align. A relatively simple fixed-charge discount should not present a problem for most billing systems, whereas a program that uses tiered benefits or tiered rates and a volumetric discount for eligible customers may require a more sophisticated billing system and a higher level of program support. Another consideration for the billing system (and program administration) is the frequency of eligibility validation for participating customers and associated updates that need to occur to keep customer information up to date.

4.3.3 Education and Outreach

For assistance programs to meet their desired goals, eligible customers must first be made aware of the program and then actively choose to enroll. To facilitate this, an active education and outreach plan will be needed to support the assistance program. This may include preparation of program communications (bill inserts/direct mail/social media) and potential partnerships with collaborators. The level of customer engagement in the program will depend on many variables, including the level of marketing and promotion. It can take significant time and effort for awareness of a new program to spread and experience from similar other assistance programs (e.g., water, energy, and other utilities) indicates that participation rates of 10% to 30% of eligible customers are typical.

4.3.4 Program Feedback/Evaluation

After launch, the utility should plan to evaluate the effectiveness of the assistance program through a continuous feedback loop. This would involve measuring performance against established metrics (looking at, for example, program participation, level of delinquencies, disconnections, program cost, participant feedback) and modifying the program as needed to achieve program goals. Collaboration with other utilities offering similar assistance programs within the service area can also be beneficial to share lessons learned and optimize the programs. This should be done alongside broader monitoring of affordability concerns within the community, as discussed in Section 6.0.

5.0 LEGAL AND ETHICAL CONSIDERATIONS

Affordability concerns are more acute today than they have been in the past, and legal frameworks have developed differently from one state to the next. Legal requirements and allowable program types vary by state, as well as between public and investor-owned utilities. The primary concern is whether revenue collected from user rates can fund the program, creating a scheme where some customers involuntarily subsidize others. Some states expressly authorize the existence of rate-funded affordability programs, while others expressly restrict it. Many states are in between, with no language or unclear language on the matter. Typically, it is considered acceptable to fund assistance programs with nonrate revenue such as voluntary contributions (e.g., bill round-up), rental income, interest, or other utility revenues.

Ethical considerations are more complex and are not the primary focus of this manual. Much has been written about the human right to water and sanitation—most notably in United Nations General Assembly Resolution 64/292 (2010)—and affordability concerns affecting environmental justice. However, wastewater utilities need to be sufficiently funded to provide an essential service, so effective affordability programs (that reach and help those who need it and only those who need it) are critical to establishing a balance between promoting affordability and continuing to provide high-quality service.

6.0 MONITORING AFFORDABILITY

Because rates and charges tend to increase over time, it is important to monitor affordability of customer bills. Only through monitoring and analysis will utility management and decision-makers be able to make informed decisions regarding operating budgets, capital plans, rate structures, and other important matters that affect the ability of the organization to fulfill their mission. However, wastewater systems seldom have the resources necessary to directly monitor and assess the affordability of rates and charges, even if a specific affordability standard has been selected. To overcome this obstacle, typically, it is possible to indirectly monitor affordability in a cost-effective manner using several types of data that most systems have available or can acquire easily.

One method to monitor affordability is to track data and trends on accounts receivable. These data could include information on the amounts of money owed for services billed and the "aging" of those billed amounts. The *aging* of accounts receivable refers to the period of time that a bill has been unpaid after the bill was sent to the customer. Accounts receivable is an indirect measure of affordability to the extent that a relatively large receivable balance or a rapidly increasing level of receivables suggests that customers are less able to pay than they may have been in the past. Similarly, if balances past due are increasing in age (i.e., more time elapsed on average since the bill was issued), then this would be another indication that overall affordability is diminishing.

Delinquent account data can also provide a good indication of affordability of user charges and related utility fees. By monitoring the number of delinquent accounts and dollar volumes associated with those accounts, utility managers can gain useful insights to the overall affordability of charges as increases in the number of delinquent accounts and related dollars owed by these accounts suggest that affordability issues may be present. Similarly, the increasing age of accounts that eventually become delinquent suggests decreasing affordability.

For those systems that have low-income assistance programs already in place, utility management can track the number of customers participating in the program, the extent of assistance being provided, and related measures of program workload performance. Changes in participation will provide insights to the ability of customers to pay their utility bills. If low-income assistance program participation increases, it might suggest that ability to pay the utility bills is dropping, unless the change is a result of increased promotion of the program to eligible nonparticipants.

In addition, local economic data that could provide insights are available from various federal, state, and local sources. Information on household incomes, unemployment rates, changes in the distribution of income, and local business license or business tax receipts can be useful in making inferences regarding the affordability of wastewater bills.

Finally, an organization can look at wastewater rates by community and compare that information to median household income by census tract (neighborhood). In completing this exercise, geographic information system personnel can cross-reference typical customer bills with income at the neighborhood level to identify areas that exceed local affordability guidelines. This results in a useful visual aid that pinpoints affordability issues in the service area. A sample output from this type of exercise is shown in Figure 12.1.

As a utility selects tracking measures that are most useful to its unique circumstances, tracking several of these types of data may help add reliability to initial conclusions on overall affordability; consistency among multiple observed trends could suggest a common cause for those trends. For example, rising bill delinquency rates, increased accounts receivable, and significantly higher unemployment rates would suggest a joint conclusion that negative local economic factors are making payment of the utility bill less affordable. Thoughtful analysis of the information that is tracked by the utility will offer helpful conclusions as to both the level of affordability and the steps that might be necessary to address ability-to-pay problems.

7.0 CONCLUSIONS

Affordability is a key factor in successfully adopting and implementing wastewater rates and charges. In many instances, it can be the most important factor in the minds of both decision-makers and the public. As a result, wastewater utilities need to carefully consider what their respective communities or individual customers within the community can afford when increases in rates and charges are being proposed and when changes in rate structures

FIGURE 12.1 Typical Wastewater Bill as a Percentage of Income

are being considered. Developing rates on a cost-of-service basis and assuming that results will be accepted, adopted, and implemented without other consideration is not viewed as practical.

The consideration of affordability should be based on two important points of view. First, the evaluation of affordability should be based on income levels, unemployment rates, or other objective standards for evaluating this issue that are relevant to the utility's circumstances. Objective standards can provide the community and its leaders with important information to make decisions on wastewater rates and charges. Second, it is often important to evaluate affordability based on community perceptions. These perceptions may significantly affect public acceptance of proposed rates and charges, which, in turn, may affect successful adoption and implementation of such proposals.

Communities should initiate and maintain means of promoting affordability through specific programs and practices directed toward this objective. These activities will help ensure that basic service is available and affordable to all segments of the local community, necessary rates and charges can be implemented, costs of bill collection and bad debts are minimized, and desired revenues can be collected to achieve the utility's goals and objectives.

If affordability is addressed and incorporated to how rates and charges are set, successful adoption and implementation of rates and charges will be substantially more likely and more effective. Both the utility and its customers will benefit, and, in the long term, this effectiveness will result in more efficient and effective provision of wastewater services to the community.

8.0 REFERENCES

American Water Works Association & Raftelis Financial Consultants. (2024). 2023 Water and wastewater rate survey. https://www.awwa.org/data-products/rate-survey/

Fitch Ratings, Inc. (2024). U.S. water and sewer rating criteria. www.fitchratings.com/research/us-public-finance/us-water-sewer-rating-criteria-29-02-2024

Grants for Construction of Treatment Works, 40 C.F.R. § 35.2140(i) (2003). https://www.ecfr.gov/current/title-40/chapter-I/subchapter-B/part-35/subpart-I

Massachusetts Department of Environmental Protection. (1997, August 11). *Guidance for abatement of pollution from CSO discharges.*

Ohio Public Works Commission. (1999, April). *Application and instructions for financial assistance* (Attachment. Water and sewer affordability standards checklist).

Teodoro, M. (2024, May 3). Good, bad, and ugly: Affordability in America, 2023 (Part 3). https://mannyteodoro.com/?p=4645

United Nations General Assembly. (2010). The human right to water and sanitation (UN Doc A/RES/64/292).

U.S. Census Bureau. (2024). *Income in the United States: 2023* (Table A-1 Income summary measures by selected characteristics: 2022 and 2023). https://www.census.gov/data/tables/2024/demo/income-poverty/p60-282.html

U.S. Environmental Protection Agency. (1995). *Interim economic guidance for water quality standards workbook* (Section 2.3).

U.S. Environmental Protection Agency. (2021). *2021 Financial capability assessment guidance*. Office of Water, Office of Wastewater Management.

U.S. Environmental Protection Agency. (2024). *Clean watershed needs survey report to Congress*. (https://www.epa.gov/cwns/clean-watersheds-needs-survey-cwns-2022-report-and-data)

U.S. Environmental Protection Agency Science Advisory Board. (2002, November 14). Untitled draft (p 4). U.S. Environmental Protection Agency.

Water Research Foundation. (2017). *Customer assistance programs for multi-family residential and other hard to reach customers* (WRF 4557).

9.0 SUGGESTED READINGS

American Water Works Association & Raftelis Financial Consultants. (2023). *Thinking outside the bill: a new guide to affordability and customer assistance*.

Announcement of small system compliance technology lists for existing national primary drinking water regulations and findings concerning variance technologies (Section A3; National-level affordability criteria), 63 F.R. 42046 (August 6, 1998).

Announcement of small system compliance technology lists for existing national primary drinking water regulations and findings concerning variance technologies (Section A3, National-level affordability criteria), 67 F.R. 62718 (August 20, 2002).

Saunders, M., Kimmel, P., Spade, M., & Brockway, N. (1998). *Water affordability programs* (p xxvii and 51). National Consumer Law Center. American Water Works Association Research Foundation & American Water Works Association.

U.S. Congressional Budget Office. (2002, November). *Future investment in drinking water and wastewater infrastructure* (Table S-2). https://www

.cbo.gov/sites/default/files/cbofiles/ftpdocs/39xx/doc3983/11-18-watersys tems.pdf

U.S. Environmental Protection Agency. (1998, February). *Information for States on Developing Affordability Criteria for Drinking Water* (Appendix F).

U.S. Environmental Protection Agency. (1998). *National level affordability criteria under the 1996 amendments to the Safe Drinking Water Act.* Prepared by International Consultants, Inc., Hagler Bailly Services, Inc., & Janice A. Beecher.

U.S. Environmental Protection Agency. (2016). *Drinking water and wastewater utility customer assistance programs.*

U.S. Small Business Administration. (1999, June 4). *Report of the Small Business Advocacy Review Panel on EPA's planned proposal of the National Primary Drinking Water Regulation for Arsenic* (Section 9.2.2). https://www.epa.gov/sites/production/files/2015-06/documents/report-sbarpanel_arsenic.pdf

Water Environment Federation. (2022). *Affordability of wastewater service* (2nd ed.).

13

Rates for Reuse or Reclaimed Water

1.0 INTRODUCTION

Reclaimed, recycled, or reused water is wastewater that has been treated for beneficial reuse following standards set by health departments or other environmental regulatory agencies. Nature has been recycling our wastewater for millions of years through a long process of sedimentation, organic consumption, natural filtration, and disinfection. Modern technology in the wastewater industry essentially accelerates this cleansing process using sedimentation, oxidation, filtration, and disinfection.

Water reuse has been practiced for decades in the arid regions of the United States and is becoming increasingly common in other parts of the country. For example, in California, Arizona, Texas, and many parts of Florida, regulations require the use of reclaimed water for irrigation if it is available and reasonably priced. Reclaimed water may offer the following advantages:

- It can be a reliable source of water. For example, in times of drought when restrictions are placed on the use of potable water for nonessential uses, reclaimed water may still be available. During periods of mandatory cutbacks and water rationing, reclaimed water can save landscaping investments, crops, commercial nurseries, and pastures. Additionally, recycled water production is relatively consistent even during periods of drought.
- If available close by, it may be a lower-cost source of water for nonpotable uses.
- Most reclaimed water still contains some of the nutrients found in wastewater and, therefore, can reduce the need for fertilizer when used for irrigation purposes.
- Large-scale reuse projects can help diversify existing water supply portfolios and reduce an agency's exposure to drought and supply disruptions. With droughts in the United States in recent history, reclaimed water has become a strategic sustainability resource.
- Reclaimed water can help preserve or enhance natural wetland habitats and groundwater basins.

2.0 FOR WHAT IS RECLAIMED WATER USED?

Reclaimed water is often used for irrigation of both edible and inedible crops, golf courses, school grounds, parks, and landscaping of public facilities and rights of way. It is also used in industry for cooling, heating, fabric dyeing, paper production, and other processes. Additionally, reclaimed water distribution systems are becoming more common in both commercial and residential areas, allowing business owners and residents to irrigate with reclaimed water rather than potable water. In addition to irrigation uses, reclaimed water can be used in snowmaking, laundry facilities, car washes, grading and dust control at construction sites, toilet/urinal flushing, and fire protection.

As the availability of water supplies from traditional sources continues to decline because of factors such as increasing population and climate change, many agencies are turning to nontraditional sources of supply such as indirect potable reuse (IPR) and direct potable reuse (DPR). Indirect potable reuse is reclaimed water that, after treatment, has been injected or percolated to an environmental barrier such as a groundwater basin, surface reservoir, lake, river, or wetland. Indirect potable reuse water is further treated through surface filtration, mixing, or ground filtration before entering a treatment facility or potable system. Direct potable reuse is reclaimed water that is highly treated and is added directly to a treatment facility or potable system without residence time in an environmental barrier. The industry has worked hard in recent years to overcome regulatory obstacles and public perceptions related to implementing IPR and DPR projects and, as a result, they are being considered to augment water supply.

3.0 RECLAIMED WATER: WATER SUPPLY OR WASTEWATER DISPOSAL?

Wastewater treatment results in the production of an effluent that needs to be disposed. Effluent may be discharged to a nearby creek, river, or ocean or through percolation ponds, per National Pollutant Discharge Elimination System or other similar state-specific permitting process. Because of regulatory requirements, outfall capacity problems or other environmental or practical limitations, it may be more economical to treat wastewater to produce reclaimed water for beneficial use than dispose of effluent by other means. Alternatively, reclaimed water may be produced to supplement potable water sources where the marginal cost of securing additional potable water may be higher than the cost of producing reclaimed water, where an agency desires to have greater control over a reliable source of water (such as the case with allocated water rights), or where an agency desires to use reclaimed water for beneficial purposes. It is important to distinguish between the two reasons for producing reclaimed water—wastewater disposal or water supply—because it may affect how the costs of producing the reclaimed water are determined and allocated.

When reclaimed water is produced because of discharge limitations on the wastewater effluent, costs associated with reclaimed water production may be passed on to wastewater customers in the form of higher sewer bills. If effluent disposal is a problem, even the costs of distributing reclaimed water may be passed on to wastewater users, with reclaimed water service being subsidized to provide greater incentives for reclaimed water use. Any income from the sale of reclaimed water may be used to offset costs to wastewater customers. In this case, there is greater incentive to provide discounts on the price of reclaimed water to encourage reclaimed water sales as a means of disposing the wastewater effluent. In some cases, reclaimed water may even be given away at no cost to the reclaimed water user. However, in the current climate, especially in the Southwest, reclaimed water is becoming a valuable supplemental source less likely to be given away.

When reclaimed water is produced to provide a supplemental source of water, the costs of producing and distributing reclaimed water may be borne by reclaimed water customers. Typically, not all of the costs of wastewater treatment are included in the production costs of reclaimed water; only the incremental costs associated with producing higher-quality reclaimed water, beyond those required for disposal actions, should be included in the production costs of reclaimed water service. When cash-flow requirements for the recovery of capital costs of reclaimed water distribution systems cause the costs of reclaimed water service to be higher than the cost of existing potable water service, water customers may be required to make up the difference between costs and reclaimed water sales revenue through higher water rates. Once capital costs have been recovered, the cost of delivering reclaimed water may drop below the cost of potable water.

To justify cost-sharing by potable water users, a comparison of the cost to potentially modify existing water systems for increased capacity should be compared to the ability to provide a reclaimed system to meet reuse water demands. Increasing water system capacity may include water resource recovery facility (WRRF) upgrades, system expansion and upgrades, and the ability and cost to obtain source water "rights" versus providing expanded capacity through reuse. When reclaimed water capacity is developed as part of an overall water resources plan, subsidy of reclaimed water by potable water and wastewater customers in the system most often can be readily justified.

4.0 PRICING ISSUES

A reclaimed water system consists of reclaimed water production facilities or advanced water treatment facilities, pumping stations, transmission lines, storage facilities, and distribution lines. The costs will vary depending on specific system characteristics. The methodologies used in pricing reclaimed water service are similar to those used for pricing water and, typically, they include a utility-specific cost-of-service study dependent on system characteristics. However, pricing reclaimed water strictly on cost of the reclaimed water service itself is complicated by several factors, as described in the following section.

5.0 COST OF RECLAIMED WATER

An issue that may arise is the ownership of wastewater effluent. In other words, who owns the water and does the treated wastewater have value? Should the costs of treatment be included in the price of reclaimed water?

These issues are relevant when the wastewater service provider and reclaimed water service provider are different agencies. As discussed previously, many of the reclaimed water service costs may be borne by wastewater customers if it is necessary to produce and dispose of reclaimed water because of wastewater discharge requirements. However, if reclaimed water is produced to supplement potable water, the incremental costs associated with production of reclaimed water in excess of the traditional treatment costs incurred to meet waste discharge requirements may be recovered from reclaimed water or potable water users. These incremental costs may include secondary, tertiary, and other treatment costs. If the wastewater service provider and potable water service provider are different agencies, depending on water market conditions, the reclaimed water may be sold to the water provider at a nominal or negotiated price. In this case, typically, the additional reclaimed water facilities (not part of the required WRRFs) are the financial responsibility of the water provider, and the costs of these additional facilities and operations are recovered from reclaimed water users and potable water users if the revenues from reclaimed water users do not meet the additional reclaimed water facilities' revenue requirements. For IPR and DPR projects, the costs of producing or purchasing reclaimed water may be considered similar to the cost other source water purchases and may be blended into existing potable rate structures, rather than being considered separately.

6.0 RECLAIMED WATER QUALITY

Historically, reclaimed water was perceived to be of lower quality than potable water and regulatory agencies did not allow direct potable use of reclaimed water. However, because of technological advances and changes in treatment processes, reclaimed water can meet and even exceed state and federal drinking standards. The quality of reclaimed water depends on the agency and the treatment processes being used. In some communities, the total dissolved solids (TDS) of the reclaimed water are relatively high, and some irrigation customers may be hesitant to use only reclaimed water. The higher TDS of reclaimed water could result in a buildup of salts and require periodic flushing with better quality potable water, thus necessitating dual connections on site (discussed in the following section). To address reclaimed water quality concerns, more advanced treatment may be necessary, which could significantly increase the costs of producing reclaimed water. Thus, pricing reclaimed water at cost, if that cost is higher than the price of potable water, would be met with market resistance. If reclaimed water cannot be priced to recover costs, then it needs to be subsidized. A significant issue in pricing reclaimed water is who should subsidize reclaimed water service.

7.0 RETROFITTING

Except in newly developed communities, typically, reclaimed water is added after the potable system has been in the ground and the plumbing at the user site has already been installed. Reclaimed water service, therefore, requires additional plumbing or retrofitting of the existing system. Backflow prevention devices are required where there is a potential for cross-contamination of potable water service. The installation of these devices is an additional cost incurred to provide reclaimed water service. To provide incentives to the user to install these facilities, even where it is mandated that reclaimed water be used where it is available, it makes sense, both from a customer-satisfaction point of view and politically to provide reclaimed water at a discount to potable water. Some agencies may install facilities for customers and recover the costs over time.

8.0 NUTRIENT VALUE

Irrigation customers (including golf courses; parks; schools; cemeteries; freeway medians; commercial and industrial grounds; and, increasingly, residential irrigation/landscape) are often significant users of reclaimed water. Because treated wastewater still contains nitrates and some phosphorus, it provides some nutrient value to plants or landscapes it is used to irrigate. The fertilizer demand is, therefore, reduced, and this represents a savings to the user. Because industrial users may not see any benefit from the nutrients present in reclaimed water, the issue is "should there be a price differential for water sold for different uses"? This is a case where the value of the reclaimed water is different for different uses. In a cost-of-service approach, there is no justification for charging differential rates for the nutrient value though, under the market-value approach, it makes sense to have differential rates. Ultimately, this is a question where simplicity of rate implementation becomes an issue.

9.0 RELIABILITY

A selling point for customers in arid regions is the reliability of reclaimed water supply. During times of drought when adequate potable water supplies are not available, irrigation customers often face maximum cutbacks to ease the burden on other customers. Because the expected supply of reclaimed water is not significantly affected by drought, a reclaimed water customer may continue to receive service without a significant cutback. This reliability issue should not be confused with the availability of reclaimed water supply on peak summer days. To maximize use of the reclaimed water system during the off-season, agencies may design a system that needs supplemental potable water or nonpotable supply to shave the summer peaks. Alternatively, excess reclaimed water from the off-season may be stored to help offset peak requirements. Typically, the cost of such storage is significant and needs to be allocated appropriately.

10.0 MARKET VALUE

Economic theory states that a commodity cannot be sold for more than what the market is ready to pay for it. The decision to produce reclaimed water should include an evaluation of economics, reliability, and control issues. Will the market pay a premium for reliability on a commodity such as reclaimed water? As long as the perception of inferior quality remains, reclaimed water cannot be completely substituted for potable water. Also, because there is often an added cost to the end user from retrofitting or installing dual plumbing systems for nonpotable uses, the price of potable water represents the upper cap on reclaimed water in a given market. Because the cost of reclaimed water, based on revenue requirements that recover the capital costs of new distribution facilities, typically is higher than that of potable water, market value plays an important part in designing reclaimed water rates.

For IPR and DPR projects, the market value of reclaimed water may not be as simple. While production costs for IPR or DPR projects may exceed the cost of alternative water sources, the reliability of the reclaimed water source may provide significant benefits. During times of drought, alternative water supplies may not be available or may be insufficient for system demands, making reclaimed water the only potential water source, increasing its overall value in the system. Additionally, when treated to potable water standards, IPR and DPR water can be used in the conventional potable water infrastructure, avoiding dual piping and plumbing systems and saving significant capital investment.

11.0 PEAKING FACTORS

Nonpotable reclaimed water users fall into two types: irrigators and industrial users. The peaking characteristics of these two types of users can vary significantly. Irrigators tend to have higher peaking factors. Their watering cycle is more intense (i.e., they use large quantities of water during shorter time frames compared to industrial users, who have more uniform daily and year-round use). Distribution facilities such as pipelines, pumping station equipment, and storage have to be sized larger for higher peaking customers. Rates can be tied to peaking factors using established methodology discussed in Section 14.4, Cost Allocation. Irrigators also tend to have much lower use in wintertime when the demand for water is reduced. In general, because a large percentage of reclaimed water is used for irrigation during the summer months, seasonal rates to stimulate demand in the winter may be considered. The peaking characteristics of different customer types may be used to differentiate rates.

For potable uses, typically, IPR users can handle more variability in flowrates as the environmental buffer also helps mitigate potential swings in water demands. Direct potable reuse users, on the other hand, may require more peaking factor considerations, depending on baseline system demands and the availability of other water sources.

Reclaimed water production remains somewhat uniform throughout the year; however, irrigation demand is high during summer and may sometimes need to be supplemented with other water sources, including potable water. This requires that reclaimed water be stored during the off-season for use in summer, or wastewater effluent needs to be disposed or stored in case of zero-discharge requirements. Typically, costs associated with such effluent disposal or storage would be borne by wastewater customers.

In addition to seasonal demands, irrigation use, especially for public use areas, may be restricted to evenings or nights, when diurnal recycled water production is at a minimum. This may require diurnal storage either through flow equalization facilities within the wastewater treatment process or through storage in the distribution system. Costs associated with distribution system storage, typically, would be borne by reclaimed water customers. Costs for flow-equalization facilities may be borne by reclaimed water and/or wastewater customers, as these facilities not only provide reclaimed water benefits but may also improve the overall wastewater treatment process.

Large commercial, industrial, institutional customers may be able to better handle nondiscounted pricing (equal to potable) of reclaimed water, especially if there is a limit on potable water use (during peak times) but no restriction on reclaimed water use. Pricing could be adjusted accordingly, depending on demand characteristics.

12.0 PUBLIC OPINION

An important part of developing a reclaimed water system is public education. Indeed, there has historically been a stigma associated with treated wastewater reuse. However, the public's commonly expressed negative view of reclaimed water can be changed and the beneficial environmental effects can be emphasized, as shown in recent years. Some educated consumers may not mind using reclaimed water at an equivalent or even slightly higher price if they feel that it is the "right thing to do" for the environment, community, and so on.

Several parameters influence the price of reclaimed water. Some parameters tend to increase the cost and others decrease it. Policymakers ultimately make decisions regarding pricing. However, it is up to the rate designer to provide different options and economics so that decision-makers can choose the option that best benefits their community. Regulatory requirements, costs, and availability remain critical factors in the ability to sell or even give away reclaimed water.

13.0 RATE-MAKING APPROACHES

During the past half century, cost-of-service, also called *cost-causative*, principles have become the predominant pricing methodology adopted for wastewater and water utility services in the United States. It appears that reclaimed water pricing will also increasingly follow this model. Subsequent to the pricing methodology, other factors will need to be considered in the rate-design phase.

The cost-causative approach used for wastewater rate-making may be used to price reclaimed water. Additionally, the cost centers are similar to those for pricing potable water. Whereas, typically, wastewater allocations are based on flow, strength, and sometimes customer parameters, reclaimed water allocations may be based on service characteristics such as average and peaking demands or on commodity and capacity demands. Either method should also recognize customer costs. In addition, as discussed previously, implementation of cost-of-service rates often requires recognition of complicated rate-design factors. For simplicity reasons, allocations may be based only on commodity or water use. Where the cost of providing reclaimed service exceeds the cost of potable water service, rates may be set by using a percentage of the potable water service pricing, an avoided cost method, or other methods. Cost-of-service rate determination is discussed in Section 14.0.

For purposes of cost-of-service rate-making, reclaimed water service is treated as being similar to water service, except that customers are limited and, typically, may have similar use characteristics, especially if they are limited to irrigation customers. *Principles of Water Rates, Fees and Charges* (American Water Works Association, 2017) covers different types of rates and approaches to rate-making. A summary of the rate-making approach and rate design for reclaimed water is covered here for completeness.

14.0 RATE APPROACHES

There are several steps involved in a cost-of-service rate-making approach. Although these steps are described in greater detail in other chapters, a brief overview is presented here for ease of understanding. They are as follows:

1. Compile user, use, and financial information regarding capital and operation and maintenance (O&M) expenses;
2. Organize data in a manner that lends itself to cost allocation;
3. Determine revenue requirements;
4. Perform cost allocation to appropriate cost centers; and
5. Design rate structure.

Each of these steps is summarized in the following sections.

14.1 Compile Data

The first step involves collection of all data required to design rates. This includes data on the number of accounts, differentiation of types of users if rates are to be developed by class, number of meters by size, use characteristics such as average and peak use, pressure zones, and so on. Users may be classified as industrial, commercial, irrigation, or landscape types and, in some cases, even residential. On the financial side, expenses of production or purchase of reclaimed water, O&M costs, and capital costs are all identified and compiled. Production costs include either the purchase cost or marginal costs of wastewater treatment that exceed treatment costs required to meet wastewater discharge requirements. Debt service costs used to finance construction of newer reclaimed water systems may represent a significant expense. Administrative costs, including overhead, customer service costs, billing, and collection costs, should be included with any required reserves. Often, a single enterprise provides water and wastewater and reclaimed service. In that case, an appropriate assignment of administrative and overhead costs to reclaimed water service should be performed. A more detailed discussion regarding financial management and accounting is included in Chapter 3.

14.2 Organize Data

To perform cost-of-service allocations, data need to be organized so that allocations can be performed readily. In any type of water service, this involves separation of production costs, including purchases of water and pumping power costs, labor, overhead, and materials and supplies associated with maintenance of production equipment. Similarly, transmission costs; distribution costs; storage costs; customer service, billing, and collection costs; and meter service costs should be defined. Capital costs may be similarly organized. For newer systems, capital costs may be amortized because, typically, debt service terms are shorter than the expected life of the system; cash-basis revenue requirements will be higher for newer systems. A more detailed discussion regarding financial management and accounting and costing for capital improvements is included in Chapters 3 and 4, respectively.

14.3 Determine User-Charge Revenue Requirements

Revenue requirements to be met by reclaimed water rates must be identified to set rates. Revenue requirements are all expenses less any revenues obtained from nonrate sources. Examples of such nonrate revenues may be interest earnings, miscellaneous fees and charges, and property taxes (many special districts collect property taxes). A more detailed discussion on determining revenue requirements is included in Chapter 5.

14.4 Cost Allocation

Allocation of costs for reclaimed water is different from allocation of costs for wastewater (discussed in Chapter 6) and similar to allocation of costs for potable water. To determine rates, costs need to be assigned to appropriate cost centers depending on the type of cost. For potable or reclaimed water service, costs may be allocated to the following costs centers (details of this allocation may be found in *Principles of Water Rates, Fees and Charges* [American Water Works Association, 2017]):

- Base costs related to annual reclaimed water demand;
- Extra capacity costs related to peaking demand over base demand;
- Commodity costs related to production of reclaimed water;
- Demand costs related to meeting maximum-day and maximum-hour peaking demands; and
- Customer service, billing, collection, and meter-servicing costs.

Commodity and demand costs may be included in the final commodity rate and customer service costs and meter-servicing costs may be included in the fixed service charge. Some agencies may include demand costs as part of the fixed service charge based on meter capacity. Demand and meter-servicing costs may vary by meter size.

In some systems, only commodity rates are charged, and these are determined by spreading the costs of the reclaimed water system over total reclaimed water sales. This rate design may work satisfactorily, especially if the users' use characteristics are homogenous. Where cost of service exceeds the cost of potable water, typically, rates are set as a percentage of potable water rates. If wastewater effluent disposal is a prime consideration, the avoided cost of the current supply becomes the upper limit of the cost to an existing user.

15.0 RATE STRUCTURE DESIGN

Many factors must be considered before development of a rate structure. The final choice often requires a balancing act among several conflicting goals. Fairness and equity, simplicity and ease of understanding, administration and implementation, incentive for conservation and effects on customers, revenue stability, and legality all must be considered before the final rate structure is adopted.

Reclaimed water rate design may be performed in the same manner as that for potable water. A survey conducted by Cuthbert and Hajnosz (1999) showed a wide range of rate structures. These included flat monthly charges, commodity-based rates, base plus commodity rates, seasonal rates, declining block rates, increasing block rates, time-of-day-based rates, take- or pay-based contracts, customer-specified negotiated rates, and separate commodity and capital charges. The survey indicated that almost 50% of utilities based their rates on potable water rates. Where service is metered, the rate structure may consist of two components: a fixed service charge and a variable commodity rate. Where service is not metered, a flat charge may be assessed based on the expected use or other consideration.

Many of the same concepts and processes for potable water rate design and, thus, potentially reclaimed water rate design, are similar to those used for wastewater rate design and are detailed in Chapter 8. Differences between water and wastewater rate design mostly relate to volumetric rate pricing approaches, which for water tend to be linked to peaking characteristics and are often designed using tiering and/or conservation-driven structures more so than wastewater volumetric pricing.

16.0 FIXED SERVICE CHARGE

Meter servicing, billing, collection, and customer service costs are recovered through the fixed service charge. In addition, a portion of O&M expenses and capital costs or debt service costs related to peaking demand may be included in this charge. Billing, collection, and customer service costs are the same for each customer and independent of meter size. Demand costs, if they are included in the fixed service charge, and meter-servicing costs may vary by meter size.

If the provider is the potable water service purveyor, to maintain uniformity with potable water service, the fixed service charge may be kept the same as for potable service and the balance of the costs to be recovered are used to determine the commodity rates. In the case where retrofitting for reclaimed water service requires dual service because the user continues to use potable water and reclaimed water, it would be unfair to charge the user the full fixed service cost on both meters. Either the fixed service charge should be maintained only for the potable service based on the old meter size or a revised (reduced) potable service meter size should be based on the potable demand characteristics of the user and billed accordingly for potable and reclaimed service. Because it may not be cost-effective to change the potable meter size, implementation of this type of pricing system becomes difficult. Retrofitted customers are then not charged a fixed service charge for the reclaimed water connection and pay only the commodity charge.

17.0 COMMODITY RATES

The variable charge per unit of consumption is the commodity rate. The balance of costs not included in the fixed service charge is recovered from commodity rates. These costs include the cost of purchasing or producing reclaimed water and a portion of the O&M, capital, and debt service costs. These costs are spread over total sales to determine the average uniform commodity rate. The rate may be modified to a declining block or increasing block rate.

17.1 Traditional Commodity Rates

Uniform rates, as the name suggests, are a constant unit price regardless of use. Most reclaimed water rate structures are uniform because they are simple to implement and understand and, typically, the customer mix is more homogenous than in potable water service.

Under the declining block rate structure, rates decrease with an increase in consumption. The rationale for declining rates is that incremental costs of providing service for increasing use are lower. Declining block rates are not common for new rate structures, especially if water supplies are more limited. For agencies that produce reclaimed water to meet wastewater-discharge requirements, it may make sense to use declining rates to provide an economic incentive to large users to switch to reclaimed water.

Increasing or inverted block rates increase with water consumption. They are used to send a stronger price signal for conservation. They can be designed to recover class-specific costs. If the user base consists of industrial customers in addition to irrigation-type customers, there may be an advantage to this rate structure. An incentive to use reclaimed water would be the availability of the water without restriction during peaks and restricted-use periods. This would also justify an increased rate with demand during peak periods but could make the rate structure more complicated to implement.

17.2 Percentage of Water Price

When the cost of reclaimed water service is more than the cost of potable water service, which happens frequently with new systems recovering capital costs on a cash-flow basis, then market economics dictate that reclaimed water be provided for less than potable water. When the potable and reclaimed water service provider is the same entity, a common practice is to price reclaimed water at a percentage of the potable water price. This discount is offered on the variable or commodity rate and not the fixed service charge, which, for simplicity, typically, is the same for potable or reclaimed water. Reclaimed water may be priced at 50% to 100% of the potable water rate and sometimes lower. Many agencies use 70% to 90% of the potable water rate. The discount is driven by market considerations to provide incentives to users to switch to reclaimed water, thus releasing demand on potable supplies and help users recover costs associated with on-site improvements needed to use reclaimed water. The balance of costs of delivering reclaimed water, typically, is picked up by the potable water users. In cases where reclaimed water is produced as a requirement of wastewater treatment, as discussed earlier, production costs of reclaimed water are picked up by wastewater users. If disposal of effluent is still an issue, reclaimed water may be given away at no cost or at a nominal price. The balance of the costs not recovered from reclaimed water use is recovered from wastewater users under the premise that those production and disposal costs would need to be incurred anyway if there were no sales of reclaimed water.

17.3 Negotiated Rates

Sometimes a user may have enough leverage to obtain a negotiated rate for reclaimed water. Typically, this happens if the user already has an independent source of water, such as well water. A reclaimed service provider cannot expect such a user to pay more than what it currently costs that user (avoided cost). Typically, such users are larger customers (e.g., power plants or industrial customers, such as paper manufacturers) who may even use all or a majority of the reclaimed water produced. An agency could sell the reliability element of reclaimed water to negotiate a deal with potential users. Quality of water may also be used as a basis to negotiate a better deal if the quality of the reclaimed water is better than the quality of the water currently available to the user.

18.0 SYSTEM DEVELOPMENT CHARGE

A detailed discussion of system development charges (SDCs) is included in Chapter 10 of this manual. A brief discussion of issues relevant to reclaimed water is presented here.

The issue of whether users should pay SDCs for reclaimed water service is often raised. Again, the issue of why reclaimed water is produced comes up. Is it to meet wastewater discharge requirements or to supplement potable water? The cost of reclamation facilities should be considered in determining the SDC for all water or wastewater users in recognition of the fact that the reclaimed water service is releasing demand on the potable water service or providing a means of disposing treated wastewater effluent, or both. The costs of facilities needed to produce wastewater effluent to meet discharge requirements should be included in wastewater SDCs. Any excess costs over and above these may be used to determine SDCs for reclaimed water or potable water. If reclaimed water is used as a supplemental water source, then, as part of a comprehensive water-resources plan, incremental wastewater treatment costs associated with producing reclaimed water and the costs of providing potable water service can be included in water SDCs for new users. An existing potable water user that releases demand on the potable system cannot be expected to pay SDCs because they have already paid potable water SDCs. However, a new water (potable or reclaimed) user should pay the consolidated SDC on both potable and reclaimed water connections. Where a customer has both potable and reclaimed water connections, care must be taken not to charge the user more than they would be charged for only a single connection. This may occur because, often, water SDCs are based on meter size and SDCs for potable and reclaimed water meters may be in excess of SDCs computed for a single, larger potable-water-only service connection. Where reclaimed water is provided for newly developing areas, the cost of infrastructure installation could be included with the cost of the development and then contributed to the agency. This could potentially reduce the agency's investment in the cost of infrastructure and make a project more feasible from a cash-flow perspective to existing users. The developer, in turn, may pay a reduced SDC based on the equivalent cost of facilities included in SDCs.

19.0 CONCLUSION

Reclaimed water can be priced based on cost-of-service principles. In many cases, because there is a dichotomy in the production and sale of reclaimed water between water and wastewater departments and disagreements in sharing of costs between them, complications may arise in pricing reclaimed water. However, this can be managed through a thoughtful, rational, and methodical process of identifying beneficiaries. It is good practice to identify the true cost of reclaimed water even though, ultimately, that cost may not be passed on to the reclaimed water customer as it may be more expensive than alternative water sources. Any revenue requirement and pricing approach for reclaimed water should be subject to all of the considerations under Section 11.0 Legal Guidelines in Chapter 6 and at least meet all applicable, local, state, and federal laws.

20.0 REFERENCES

American Water Works Association. (2017). *Principles of water rates, fees and charges* (Manual M1, 7th ed.).

Cuthbert, R. W., & Hajnosz, A. M. (1999). Setting reclaimed water rates. *Journal of the American Water Works Association, 91*(8), 51–57.

14

Emerging Financing Alternatives

1.0 INTRODUCTION

As discussed in Chapter 4, typically, wastewater system financing involves a combination of external- and internal-funding sources. Internal sources are those where funds are provided through collections of utility revenues (i.e., rates, charges, special assessments, and dedicated taxes) or derived through cost reductions or depreciation. External funding involves accessing other entities to raise funds, whether through capital markets in the case of debt issuance, the development community in the case of system-development charges and contributions, private- or public-sector entities in the case of partnership arrangements, or governmental entities in the case of grants and state revolving fund loans.

Wastewater utilities have a long and staid tradition of financing their systems through these respective sources. For the majority of systems, user charges are the primary source of revenues. This is especially true in the United States, as federal support for wastewater system development through construction grants has waned in favor of low-interest state revolving loans supported by utility revenue pledges, and various forms of taxation have been challenged and restricted through voter initiatives (Pew Charitable Trusts, http://www.pewtrusts.org/en/multimedia/data-visualizations/2014/fiscal-50#ind0). This reliance on internal sources of funding as the foundation for utility financing has resulted in user-charge increases more than general inflation indices as utilities address increasingly stringent regulations as well as asset renewal and reinvestment backlogs (National Association of Clean Water Agencies, 2018). These trends have occasioned the emergence of a variety of innovative financing approaches that leverage new technologies, special debt instruments, and new partnership structures to mitigate rate pressures.

This chapter provides a review of the most prominent emerging financing alternatives that have characterized the wastewater industry. These include funding options that

- Use new technologies to manage operation and maintenance (O&M) expenses,
- Efficiently deliver capital projects and manage associated risks,
- Exploit resource recovery opportunities to reduce expenses and establish new revenue streams,
- Leverage options to tailor debt obligations, and
- Establish new partnership structures.

2.0 NEW OPERATION AND MAINTENANCE EXPENSE MANAGEMENT STRATEGIES

Various wastewater utility business practice and technology advances have brought forward a wide range of opportunities to realize significant improvements in operating efficiency. These include but are not limited to the following:

- Treatment process optimization studies that often identify means to meet or exceed water-quality regulations while reducing use of costly chemicals and electricity.
- Instrumentation and control systems that improve monitoring of system operations and may enable reductions in staffing levels.
- Computerized maintenance management systems that may help improve the efficiency and effectiveness of field operations through measures as simple as revised routing or field crews to limit travel times between jobs.

- Cooperative procurement arrangements whereby individual utilities join to negotiate better terms for commonly used chemicals, supplies, and equipment that render lower costs for specific items. These arrangements may also help forge cooperative arrangements for storage and inventory management that yield further efficiencies.
- Energy-management programs that often reduce energy costs dramatically through installation of more energy-efficient and, often, more reliable equipment. These programs optimize energy use by taking advantage of time-of-use pricing by local energy providers. Energy-use optimization, like resource recovery, is particularly suited to a simple form of public–private partnership (discussed further in the following section) whereby utilities may enter into contracts that provide for sharing of savings realized through installation of advanced technologies.

Because O&M expenses are not amortized through financing but are paid directly as a component of system revenue requirements, O&M expense reductions provide a direct way to "free up cash" to either support capital reinvestment or mitigate rate pressures. However, typically, adopting new technologies includes implementation and annual support and maintenance costs. This emphasizes the need for wastewater utilities to perform a cost-benefit analysis before implementing technology advances.

3.0 CAPITAL PROJECT DELIVERY EFFICIENCY AND RISK MANAGEMENT

Equally important, advances in capital project delivery are increasingly enabling wastewater utilities to reduce costs and manage risks associated with implementing infrastructure investments. Cost reduction strategies are particularly important because wastewater utilities are in one of the most capital-intensive of all economic sectors (The Water Research Foundation, 2009). Recent advances include but are not limited to the following:

- Design, material, and construction technologies that enable wastewater utility contractors to more efficiently and precisely manage design processes and use higher-quality building materials and a range of advanced construction practices;
- Program controls that help ensure coordination and efficiency of various project-delivery phases across individual projects and other utilities with assets in common easement areas; and

- Life-cycle cost management that helps ensure that assets realize their intended useful lives by guiding design and material specification standards, ensuring proper O&M of built assets, and monitoring asset conditions to appropriately schedule renewal and rehabilitation measures.

Additional information on these advances can be found in various Water Environment Federation publications (2013, 2014, 2015), perhaps most notably those offering "roadmaps" to resource recovery and energy management and via industry organizations such as the Association for the Improvement of American Infrastructure and the Design–Build Institute of America.

Given the pronounced and well-documented economic costs associated with failing to replace infrastructure (American Society of Civil Engineers, 2021), wastewater utilities' adoption of these strategies to efficiently and effectively deliver its capital projects could not be more critical.

Although expense reductions, whether O&M or capital, do not provide new sources of funding per se, expense optimization does help ensure that utility revenues are not wasted and are dedicated to quality service delivery. As rate pressures continue, demonstrating stewardship of existing revenue streams will be increasingly important to help secure additional funding. Utilities are, therefore, well served to use benchmarking efforts to gauge their performance relative to their peer utilities and other entities performing similar activities and processes. In doing so, it is important that utilities selected for comparison are true peers (or appropriate adjustments are made to recognize significant differences) for valid conclusions to be drawn. Financial performance benchmarking may be particularly important for wastewater utilities accessing capital markets and seeking credit on favorable terms. Typical metrics (National Association of Clean Water Agencies, 2018) include the following:

- Budget versus actuals;
- Electricity, chemical, and laboratory costs per million gallons (m^3);
- Operation and maintenance cost per million gallons (m^3);
- Total cost per million gallons (m^3);
- Days operating cash reserve;
- Long-term debt;
- Debt service coverage; and
- Residential sewer charge.

4.0 RESOURCE RECOVERY

Although expense management has been a hallmark of effectively managed utilities and has historically been highlighted by emerging techniques, resource recovery represents an entirely new paradigm for the industry, with a host of financial-management implications. For example, traditional rate-setting practices reflect the view of wastewater system influent streams as system demands for which charges are established to reflect the costs of management and disposal. Equitable distribution of responsibilities for these costs has historically been defined by waste disposal required by individual user classes. These models do not take into account influent streams as resource inputs subject to recovery and as opportunities for revenue. Many utilities have developed resource recovery options by harvesting phosphorus for fertilizer production and/or digester gas for generation of heat or electricity.

Resource recovery options present the fundamental question of how net revenues associated with recovery operations should be treated in determining wastewater rates. Options include but are not limited to the following:

- Segregated operations. This approach effectively sets aside the question of determination of wastewater rates. These rates are determined as if no value is associated with resource recovery for purposes of wastewater rate setting. The utility obtains the resources effectively at no cost and its resource recovery business is treated as a separate enterprise.
- Systemwide revenue requirement reduction. This approach, most commonly used among utilities with existing resource recovery operations, is to regard resource recovery operations as benefitting the entire wastewater system. Net revenues accrue to the system in aggregate and effectively serve to reduce general system rate revenue requirements in much the same way as miscellaneous revenues and interest earnings. Under this approach, resource recovery operations are treated as a component of rather than separate from the wastewater system and all ratepayers benefit from monetization of recovered resources.
- Selected contributor credits to wastewater rates. Particularly if resource recovery is afforded by especially strong concentrations of contributed waste streams, there may be a basis for providing wastewater rate credits (or other forms of remuneration) in recognition of the value of received resources. This option may have appeal as a means of encouraging delivery of high-strength loads to utility receiving points (potentially as a counterpoint to enforcement actions) or in support of economic development/retention initiatives targeted to (typically)

industrial users. However, it imposes methodological complexities that must be addressed in the rate-setting process. For example, it requires a determination of strength concentration/loading levels that convey eligibility for rate credits provided and evaluation of relative cost differences involved in handling loadings from different types of users.

Beyond this fundamental issue, other rate-setting complexities derive from the practical realities of establishing a resource recovery concern. For example, to the extent that advance investments are required to implement resource recovery technologies, consideration must be given to how these investments will be financed, how any associated indebtedness will be secured, and if/how associated costs are included (even temporarily) in wastewater system revenue requirements. As a consequence, the movement to resource recovery has rate-setting implications that require substantial vetting with applicable governing boards and regulatory authorities.

Monetized value creation, which is the generation of revenues from resource recovery operations, also presents a host of additional financial-management challenges. For example, resource recovery operations that involve sales in a competitive marketplace impose requirements to assess market conditions, evaluate product/service-pricing options, and evaluate capital investments based on risks and return opportunities largely foreign in a natural monopoly setting. Additional financial reporting and analysis responsibilities are dictated by both regulatory review and accounting requirements. For circumstances where resource recovery operations are treated as functionally separate from wastewater operations, procedures may effectively mimic those of a for-profit enterprise. Financial statement equivalents may be developed to present information on book values of associated assets, profitability, and liquidity, which are all likely shown in a separate enterprise fund. Supporting information on, for example, customer accounts receivables or cost-accounting measures may go further to inform interested stakeholders. For this reporting, and perhaps more acutely for integrated operations (for which separate enterprise may not be established), clarity with regard to how shared operating and capital costs (particularly indirect charges) are allocated across functions is a fundamental concern.

5.0 DEBT INSTRUMENTS

In the same way that the wastewater utility financial management has been subject to profound changes with the advent of resource recovery, recent evolutions in the capital markets serving wastewater utilities present a much broader range of financing options than ever before. These options enable wastewater utilities to tailor their debt-repayment schedules, access new funding programs and providers, and build new partnerships. Although many wastewater utilities have and will continue to rely on a traditional debt portfolio composed of fixed-rate, long-term revenue bonds and state revolving fund loans, a raft of new options can provide savvy issuers opportunities to save on borrowing costs and effectively manage associated rate pressures. In each case, when evaluating different options for obtaining financing, differing transaction costs should be carefully evaluated and considered. Noteworthy options that are particularly well suited for wastewater utility capital program financing include the following:

- Tax-exempt commercial paper (TECP). These are short-term, typically 90- to 270-day, unsecured notes that, typically, are used by wastewater utility issuers as a "bridge" financing mechanism during capital project design and construction. Although not necessarily a new financing instrument, use of TECP in the municipal wastewater sector has increased in recent years. Their principal advantage is the flexibility they provide to utilities conducting significant capital programs. Typically, TECP notes rank subordinate lien and right of payment to revenue bonds and are often enhanced by credit facilities (e.g., letters of credit). The associated lines of credit provide funding capacity for utilities whose procurement policies require funds to be encumbered for contracts to be let. Utilities may use TECP notes to procure and fund initial project design and construction costs and then size and time issuance of long-term debt obligations that may be used to repay the outstanding notes at the utilities' discretion. This flexibility has proven to be particularly useful for significant wastewater capital program implementations, enabling utilities to right-size their debt portfolio and avoid project funding delays; however, these programs require active management because TECP notes need to be remarketed until they are refinanced through a long-term debt issue.
- Special debt types. In contrast to commercial paper instruments that have historically been used to meet short-term liquidity needs and for which there is a well-established market, in recent years a number of new debt instruments have been designed to meet evolving investor demands and/or issuer needs. Several of these instruments have particular applicability for wastewater systems, including the following:
  - Long-duration bonds. These are bonds with debt repayment periods exceeding the typical 30-year term. These multigenerational term bonds address a long-standing concern in the wastewater industry that even 30-year revenue bond issues do not adequately match

debt-repayment periods with the extended useful life of many wastewater infrastructure assets. This is particularly true for wastewater conveyance networks where tunnels and significant pipelines have useful lives typically expected to exceed 75 to 100 years. These bonds may be particularly useful for lowering annual debt service requirements (by extending the repayment period) and, under favorable market conditions, may enable wastewater utilities to lock in low-interest rates for an extended period.

- Green bonds. Green bonds are used to finance projects intended to provide significant and stipulated environmental benefits. Project types initially were focused on climate change adaptation and resiliency efforts but are expanding to a broad range of investments, including, for wastewater utilities, projects focused on water-quality improvement, resource recovery, and renewable energy development. Green bonds require issuers to make commitments to monitor and disclose the environmental performance of the financed projects, and they may be subject to penalty payments or other clauses if projects underperform. This structural feature has given rise to concerns about the standards by which environmental performance is judged and the potential for "green washing" as issuers try to take advantage of enhanced and diversified investor demand as well as the branding benefits of issuers' public commitment to the environmental performance of financed projects. In the case of wastewater utilities' green bonds, this is less of a concern because most wastewater capital projects provide clear environmental benefits.
- Social impact/environmental impact bonds. Although green bonds offer noteworthy benefits through the branding and public relations value of the designation and are publicly traded instruments, social and environmental effects tie the returns on the financial investment to documented social and/or environmental effects of the financed project. These "pay-for-success" types of instruments are not sold in publicly traded capital markets but are essentially loans provided by private parties to fund projects designed to deliver desired social or environmental effects. Typically, investors provide capital for a project or program implementation that is executed by an external organization or intermediary who has a performance-based compensation contract with a governmental organization or nongovernmental agency. Wastewater utilities could, for example, provide compensation to green infrastructure developers based on volume of stormwater captured. The utility would only pay for performance of the installed green infrastructure as verified by an independent evaluator. These types of instruments are of particular

CASE STUDY

DC Water Century Green Bonds

Issuance: July 2014
Maturity: 100 years
Amount: $350 million
Credit Rating: Aa2/AA+/AA
Project Name: DC Clean Rivers Project
Project Type: CSO tunnel
Project Cost: $2.6 billion
Project Exp. Life: 100 years

This was the first issuance of a century bond by a municipal water utility in the U.S. While DC Water (Washington, DC) had typically offered comparatively shorter-term bonds, the expected life of tunnels to be constructed as part of the Clean Rivers program provided a prime opportunity to explore a longer term. The long duration is viewed as addressing inter-generational equity by distributing the cost amongst all who will benefit from the project. The issuance also was the first to incorporate a sustainability certification process by an independent second party. The monumental project, consisting of deep tunnels to convey stormwater and wastewater has clear water quality, waterfront restoration, and climate resilience benefits. (DC Water, 2014)

CASE STUDY

Metropolitan Water Reclamation District of Greater Chicago – Green Bonds

Issuance: December 2014
Amount: $297 million
Credit Rating: AAA/AAA
Project Categories: Tunnel and Reservoir Plan (Series A bonds), Stormwater Management Program Projects (Series B bonds), Resource Recovery Projects (Series C bonds), Water Reclamation Plant Expansions and System Improvements (Series C bonds)

The Metropolitan Water Reclamation District of Greater Chicago (Chicago, IL) issued one of the early green bonds in late 2014. The Tunnel and Reservoir Plan (TARP) program, begun in 1972, focuses on CSO reduction and protection of Lake Michigan. Proceeds also funded $228 million for stormwater management projects and $118 million for biosolids management improvements. (WaterWorld, 2014)

FIGURE 14.1 Examples of Green Bonds

interest in that they engage a unique set of motivated social impact investors and tie returns to independently verified performance yet have initial, involved, daunting transaction costs that may limit widespread applicability.

- Water Infrastructure Finance and Innovation Act (WIFIA) Loans. Enacted in 2014, this act established a federal program to make available low-interest-rate loans to partially fund (up to 49%) the cost of significant water and wastewater infrastructure projects costing at least $20 million (or $5 million for communities of fewer than 25,000). Total federal funding for a project partially funded with a WIFIA loan cannot exceed 80%. The WIFIA is modelled after the Transportation Infrastructure Finance and Innovation Act program enacted in 1998 that provides "credit assistance for qualified projects of regional and national significance" (U.S. Department of Transportation, 2015). The WIFIA loans, with a maximum 35-year term, will be issued at an interest rate equal to (or marginally greater than) the U.S. Treasury rate of a similar maturity at the date of closing, providing a borrowing cost advantage even over tax-exempt municipal debt. The program also allows for payments to be deferred for up to 5 years after the project reaches substantial completion. Projects funded through WIFIA loans must comply with the National Environmental Policy Act, Davis-Bacon, American Iron and Steel, and all other federal cross-cutter provisions.

Complementing these special debt types are evolving arrangements that may address many wastewater utilities' emerging or atypical capital-financing needs. Private placements and securitizations may give utilities with limited access to credit markets an opportunity to finance projects with different lenders; catastrophe and resilience bonds may be used in conjunction with insurance coverage to manage climate change risks and other significant loss exposures. These and other debt options, discussed in greater detail in The Water Research Foundation publication titled *New and Emerging Capital Providers for Infrastructure Funding* (2016), may continue to evolve with the dynamics of the utility marketplace.

6.0 PARTNERSHIP ARRANGEMENTS: PUBLIC–PRIVATE AND PUBLIC–PUBLIC

Similarly, various forms of partnerships largely introduced to the wastewater utility marketplace over the last 20 years are evolving to meet the dynamic financing needs of the industry. Hallmark and interrelated attributes of these partnerships include the following:

- Leveraging of respective parties' financial positions, tax status, and/or access to capital resources to enable or reduce the overall costs of project financing.
- Integration of one or more traditional project-development phases (e.g., planning, design, permitting, construction, operation, financing) that alter the legal structure of the parties and financial terms of the project development.
- Contractual allocation of risks based on parties' respective abilities to manage or mitigate those risks. Capital project delivery risks include but are not limited to design quality, site conditions, supplies and materials cost escalation, construction performance, project permitting, and project-financing terms.

More comprehensive public–private partnerships (P3s) are characterized by contracts that integrate design, construction, financing, and/or O&M elements through a single contract, often over a lengthy concession term. Types of collaborative project-delivery models are construction manager at risk, Design–Bid–Build (DBB), Design–Build, Design–Build–Operate, and Design–Build–Finance–Operate.

In some cases, a special purpose vehicle (SPV) is established as a legal entity solely for a project. This structure limits the liability of investors and makes the project less susceptible to liabilities of the firms making up the SPV. Assets and liabilities of the SPV are only those related to the project. The SPV receives equity contributions from the firms that are equity members and pays dividends to investors in return.

To date, most successful P3 projects have been for larger projects where the transaction costs are limited relative to the total project cost. Many transaction costs associated with P3 projects are not proportionate to project cost, yielding economies for larger transactions. In recent years, attempts have been made to aggregate projects into packages to reduce transaction costs and make the P3 model more viable for smaller system projects (CH2M HILL, 2012).

Tax implications also must be carefully considered when choosing a project-delivery method. Although tax-exempt municipal funding is available at a lower cost and is readily accessible, benefits of the partnership model may mean that the total life-cycle cost of the project is lower despite higher capital costs. Access to tax-exempt financing through exempt facility private activity bonds (PABs) may also be a viable alternative, although an important consideration is ensuring that the tax-exempt status of the PABs is maintained throughout the project life cycle.

Frequently, public owners will complete a "value for money" analysis, which compares the P3 procurement method for a given project against a traditional DBB method, or a "public sector comparator." This comparison should reflect the total life-cycle cost of the project, taking into account efficiencies of risk allocation, to ensure valid comparisons. By steering away from comparing bid costs that may not be truly comparable and quantifying the value of risks transferred agencies can make better, more informed decisions.

In doing so, it is important that the utility's capital project planning and public engagement processes be adjusted to take into account the integrated project-delivery structure common in P3 projects. For instance, a common P3 overall "project cost" could encompass professional services for design, capital construction cost, O&M, and debt servicing for the entire concession term, which can be as long as 30 years. Resultant life-cycle pricing covering the entire 30-year period could be misconstrued by governing boards, other stakeholders, and the general public in the absence of proactive and effective communications.

6.1 Risk Allocation

Similarly, in structuring partnership arrangements, a fundamental determinant of whether the process will be tractable is the extent to which the attributes of risk and principles of risk allocation are understood by all parties. Private partners are frequently engaged to provide integrated services, in part with an expectation of their willingness to accept substantial project-development risks. Public entities that fail (or are unwilling) to recognize the constraints on risk acceptance of commercial enterprises may expect limited private sector interest in a potential partnership, strained contract relations, or private financing on less favorable terms. Time invested by public owners and private partners in a comprehensive risk-allocation exercise will pay dividends later in the project.

Although risks can never be mitigated to zero and one party cannot take all of the risk, risk allocations may promote effective risk management, mutually beneficially contract terms, and reasonable financing terms. To do so, parties involved in the partnering process must recognize the respective positions of each party and work collaboratively to find the optimum risk profile for the optimum cost, where neither too much risk is transferred to the private sector nor does the public owner retain too much risk.

6.2 Procurement Process

Creating a sound and reasonable solicitation process is key to ensuring commercial interest and competition in a P3 procurement. Of particular value can be early industry forums, frequently conducted before the start of the official procurement. Another element to consider for inclusion is time for an alternative technical concept (ATC) process. This process allows shortlisted proposers to suggest design or operational changes that may deviate from the performance-based technical specifications but offer value to the project over its life cycle. Stipend structures are frequently set up to allow the owner to compensate nonselected proposers for the intellectual property contained in their ATCs and contract provisions frequently allow the selected proposer to use these at their risk.

The P3 arrangements may yield significant value to utilities through the active participation of the private sector in structuring project financing (potentially including their equity contributions), encouraging innovation, providing flexible risk-transfer terms, and improving overall project performance. Although procurement laws vary across different states with respect to the use of alternative project delivery methods such as P3s, industry organizations track enabling legislation and can provide help with lobbying or best practices to be implemented in states that are beginning the process.

7.0 IMPLICATIONS FOR FINANCIAL PLANNING

With the array of efficiency strategies, resource recovery technologies, and partnership arrangements available to manage costs and secure financing, wastewater utilities are well served to develop long-term strategic financial plans. In general, such plans extend the forecast periods of cash-flow analyses used to demonstrate the feasibility of debt issues and may project how differing financing options affect systemwide rates and key financial-performance metrics such as debt service coverage or projected debt-to-equity ratios. Strategic financial planning may help utilities define a clear strategy for how to use the increasingly diverse array of financing options to plan rate adjustments, manage and mitigate risks, and protect their financial integrity, which, in turn, may help utilities secure system financing on favorable terms and stipulated benefits.

8.0 SUMMARY AND CONCLUSIONS

In recent years, wastewater utility financing has evolved from a relatively staid practice of balancing cash financing and traditional forms of debt, most prominently state revolving fund loans and tax-exempt municipal debt obligations. In the 21st century, wastewater utility financing will increasingly involve strategic deployment of an array of options in a dynamic market populated by new technologies, new lenders, and new partners.

9.0 REFERENCES

American Society of Civil Engineers. (2021). *Failure to act: Impacts of status quo investment across infrastructure systems*. https://infrastructurereportcard.org/the-impact/failure-to-act-report/

CH2M HILL. (2012, November). *West Coast Infrastructure Exchange final report*.

DC Water. (2014). DC Water announces successful sale of $350 million green century bonds. https://www.dcwater.com/about-dc-water/media/news/dc-water-announces-successful-sale-350-million-green-century-bonds

National Association of Clean Water Agencies. (2018, July). *Opportunities & challenges in clean water utility financing and management, financial survey executive highlights*. https://www.nacwa.org/docs/default-source/resources---public/2018-08fina-survey-summary.pdf?sfvrsn=b5acf661_4

Pew Charitable Trusts. Fiscal 50: State trends and analysis. http://www.pewtrusts.org/en/multimedia/data-visualizations/2014/fiscal-50#ind0

The Water Research Foundation. (2009). *Improving water utility capital efficiency*.

The Water Research Foundation. (2016). *New and emerging capital providers for infrastructure funding* (Project No. 4617).

U.S. Department of Transportation. (2015). Transportation Infrastructure Finance and Innovation Act of 2015. Program overview. https://www.transportation.gov/buildamerica/financing/tifia#:~:text=The%20TIFIA%20credit%20program%20offers,reasonably%20anticipated%20eligible%20project%20costs

Water Environment Federation. (2013). *The Energy roadmap: A water and wastewater utility guide to more sustainable energy management*.

Water Environment Federation. (2014). *Moving towards resource recovery facilities*.

Water Environment Federation. (2015). *The nutrient roadmap*.

Water Environment Federation. (2021). *Energy in water resource recovery facilities* (Manual of Practice No. 32, 2nd ed.).

Water World (2014, December 12). MWRD to offer investors 'green bonds' for environmentally-friendly projects. https://www.waterworld.com/drinking-water-treatment/potable-water-quality/article/16215269/mwrd-to-offer-investors-green-bonds-for-environmentally-friendly-projects

Appendix

Utility-Basis Revenue Requirements and Cost

1.0 INTRODUCTION

Most frequently, municipal utilities in the United States use the cash-needs approach for defining utility costs used in cost-of-service allocation studies because the approach parallels municipalities' budgeting process. However, many other municipal utilities use a utility-basis approach for their cost-of-service allocations for a variety of reasons.

This appendix covers the utility-basis approach for defining revenue requirements and allocating costs by cost components to customer classes served by municipally owned wastewater utilities. To allow a comparison of utility-basis approach results with cash-based results shown in this manual's

Chapters 6 and 7, this appendix will use the same case study and will point out similarities and differences.

2.0 UTILITY-BASIS APPROACH

The utility-basis approach provides an appropriate method for measuring revenue requirements and calculating costs of service applicable to all classes of customers served by a wastewater utility. The approach works particularly well for allocating costs to those customers located outside the corporate limits of a municipally owned utility. In such situations, the service relationship parallels that of an investor-owned utility because the owner (the municipality) serves non-owner customers (those outside the corporate limits).

Customers inside the municipal corporate limits may be considered to be the utility's stockholders. They are ultimately responsible for paying all operating and capital costs of the utility should outside customers decide to no longer be served by the municipality. Thus, the municipality is entitled to a reasonable return from the non-owner customers, based on the value of its assets that are used and useful in providing the service to them and the business risks inherent in municipal service to non-owner customers. As stockholders, customers inside the corporate limits are also reasonably entitled to share in the municipality's recovery of its business risks from non-owner customers through lower municipal taxes or utility bills.

Typically, outside users do not have a political voice representing their rate interests on the regulatory body for the municipally owned utility. Accordingly, the utility basis provides a measure of protection for the non-owner customers against inconsistent capital financing choices or inequitable rate-setting practices of the municipal owner. The utility-basis approach often stabilizes periodic rate changes compared with the cash-basis approach as revenue- or debt-financed capital improvements may fluctuate significantly from year to year.

Municipal utilities use the utility-basis approach for reasons other than business risk recovery, such as the following:

- Some states mandate that both municipally owned and investor-owned utilities have their rates approved by their state's public service or utility commissions.
- Some states require that municipally owned utilities providing service to entities outside their municipal corporate limits have outside rates reviewed and approved by the state's public service or utility commissions either before rates are implemented or when challenged by outside customers after they are implemented.

- Some municipally owned utilities serving outside their municipal corporate limits use the utility-basis approach to help justify the differential in rates charged customers inside and outside municipal corporation limits.
- Wholesale service agreements between municipalities may specify the utility-basis approach for rate-setting to minimize potential for variability in politically set rates.

3.0 REVENUE REQUIREMENTS DEFINITION

Table A.1 shows typical revenue requirements for utility-basis and cash-needs approaches, assuming that the utility's intent is to generate the same level of rate revenue for either basis.

Utility-basis approach revenue requirements result from the determination of operation and maintenance (O&M) expenses, annual depreciation, property taxes in various jurisdictions, or payments in lieu of taxes (PILOT), and an appropriate level of annual return on rate base. The elements parallel the statement of revenues, expenses, and changes in fund net assets (previously called income statement) in audits for municipal enterprise funds showing revenue from all sources less O&M, depreciation, and taxes (or PILOT) resulting in net income, or return, to the entity. Positive net income does not necessarily mean that the utility is financially sound. Well-financed utilities generate sufficient annual revenue to fund the following consistently:

- Annual operating costs (O&M, taxes or PILOT, and existing debt payments),
- Routine capital expenditures for existing infrastructure rehabilitation and minor capital replacements,
- Maintaining credit ratings and investor confidence in their financial integrity, and
- Needs by investor-owned utilities for attracting or retaining capital investment.

Revenue requirements for the municipal cash-basis approach consist of O&M; property taxes from outside corporate limits or PILOT from inside corporate limits; annual debt service on outstanding and new bonds used to finance significant capital expenditures; and revenue-financed routine capital expenditures for equipment, significant repairs, and minor rehabilitation projects.

TABLE A.1 Utility-Basis and Cash-Based Revenue Requirements and Return

Line	Utility basis		Cash basis	
no.	Revenue requirements		Revenue requirements	
1	Operation and maintenance expense	$18,202,000	Operation and maintenance expense	$18,202,000
2	Operation and maintenance reserve	66,000	Operation and maintenance reserve	66,000
3	Payments in lieu of taxes	975,000	Payments in lieu of taxes	975,000
4	Depreciation expense	3,814,000	Debt-service requirements	8,030,000
5	Return	5,796,000	Equipment purchases	580,000
6			Capital improvements from revenue	1,000,000
7	Total revenue requirements	$28,853,000	Total revenue requirements	$28,853,000
8	Less other revenue	(17,000)	Less other revenue	(17,000)
9	Net costs of service to be met from rates	$28,836,000	Net costs of service to be met from rates	$28,836,000
	Return development		Return development	
10	Original cost less depreciation rate base	$131,454,000	Original cost less depreciation rate base	$131,454,000
11	System return as percent of rate base	4.41%	System return as percent of rate base	4.41%
12	Outside rate of return	7.25%		
13	Inside rate of return	3.53%		

4.0 REVENUE REQUIREMENTS COMPARISON

Table A.1 also facilitates a comparison of revenue requirements for the utility- and cash-basis approaches. Under either basis, O&M and PILOT are exactly the same. In the defining of capital cost elements, the bases differ. The utility-basis approach uses the elements of depreciation and return to encompass the cash-basis elements of annual debt service, equipment purchases, and routine annual capital expenditures.

4.1 Depreciation

Depreciation represents the annual loss in service value not restored by current maintenance of depreciable utility assets resulting from the provision of service to utility customers. Among the causes are normal wear and tear; environmental action resulting in damage or material deterioration of system components; and inadequacy or obsolescence because of changes in technology, customer wastewater contributions, and regulatory agency policies and requirements. The proper level of annual depreciation expense should be based on appropriately estimated average remaining service lives and the original costs when constructed or acquired of depreciable assets currently in service.

Depreciation does not represent an actual utility annual cash outlay. However, if annual utility revenues properly cover annual depreciation, funds should be provided to meet the actual cash requirements of annual debt principal payments and a portion of routine capital expenditures. Depreciation is computed for all depreciable assets each year, but not all such assets have outstanding debt associated with them. Although there may be a significant difference in the length of the average service life for all assets and typical bond terms (or number of debt repayment years) for outstanding bonds, most utilities' capital structures result in more depreciation funds being generated each year than existing debt service annual principal payments. The resulting excess funds may be used for other capital-related expenditures.

4.2 Return

Return is intended to cover annual interest costs of the debt-financed portion of the utility's assets and a portion of routine capital costs while providing a fair rate of return on the equity-funded portion of assets. Although annual debt-interest costs are determined readily, the cost of equity capital for a utility is considerably more difficult. As a general rule, total return to the municipal owner on both debt and equity portions of assets, or total rate base, should be set in keeping with the level of return for other enterprises having corresponding risk.

The total annual return is obtained by applying the weighted rate (weighted by the proportions of debt and equity in the utility's rate base) of return to the utility's rate base. In most jurisdictions, the rate base is computed as the net book value of the facility in service and, typically, is based on originally recorded, or book, cost less an accumulated reserve for depreciation. Typically, rate base is further adjusted by allowances for cash working capital and construction work in progress. Contributions in aid of construction, including facilities donated by developers or other customer funds, are deducted from the facility in service for rate base determination. Thus, return is applied to owner-supplied facilities only, that is, to the remaining investment that the owner has at risk in utility assets.

Table A.1 shows a system rate of return of 4.41% being required to complete the utility's cash needs for the year. The system return is the break-even return required by the municipal utility to operate properly as it meets its requirements according to utility management's plans. An investor-owned utility needs a return to meet the same cash needs plus provide a dividend to its stockholders. Thus, typically, investor-owned utilities' rates of return are larger than those for municipally owned utilities.

After determining the system return, the municipality must decide how to recover the return from its customers. If the municipality serves solely within its corporate limits, likely, the utility would charge the same rate of return, that is, the system rate of return, on the rate bases serving its customer classes. This would result in total allocated costs to customer classes similar to the results using the cash-needs basis in Chapters 6 and 7. If the municipality serves retail and wholesale customers outside its corporate limits, the utility would probably assess the level of risk involved in serving outside customers and charge a higher rate of return than the rate of return for inside customers.

To derive a higher outside rate of return, typically, municipalities either add a fixed percentage in the range of 2% to 4% or apply a multiplier in the range of 1.25 to 2.00 to the system rate of return as an allowance for costs associated with risk. For municipal systems, inside-city customers' (e.g., effective "stockholders") rate of return could be reduced to the point where revenue generated in total covered the remaining amount of return not paid by outside customers; that is, the 3.53% return shown in Table A.1. Alternatively, the extra funds could be transferred to the municipality's general fund to offset the need for additional tax revenue from its residents.

Table A.1 shows an outside rate of return of 7.25%, which equates to an approximately 1.64 multiplier applied to the 4.41% system return. A multiplier this high may be indicative of a service situation where the municipal utility has no means of terminating service to outside retail customers if they fail to pay. The situation could also apply where wholesale service is provided without contract. Lower-risk multipliers may be warranted if, for example, the municipality has a contract with an outside jurisdiction to assist with bill collection for nonpaying retail customers or a wholesale contract that requires the customer to contribute an amount equivalent to its rate base should the customer elect to end its service contract. These are simply two examples of potential situations that would warrant a lower "risk factor" multiplier.

A hybrid utility- and cash-basis approach, referred to as the *utility basis with cash residual approach*, may also be used in situations where a municipally owned utility provides utility service to customers outside of its political boundaries. Total revenue requirements for the utility are established using the cash-basis approach. Revenue requirements for those customers located outside the political boundaries of the utility are determined based on the utility basis, reflecting the O&M expense, depreciation expense, and a fair return on rate base allocable to these customers. Because utility revenue must meet all budgeted cash-basis revenue requirements, revenue requirements from outside users that recognize the utility basis are deducted from the total system revenue requirement, with the cash residual being recovered from inside users.

4.3 Risk

Risk is a reality and has real costs associated with it for municipal utilities that provide service outside their corporate limits. The risk multiplier shifts utility costs from inside customers to outside customers for future additional costs that outside customers may cause the municipal utility to incur on their behalf. This payment may be equated to a periodic insurance payment to the utility where the utility is self-insured. When any of the following risks materialize, the utility will spend additional funds to address these outside-customer issues:

- Outside customers are slow in making bill payments, requiring increased collection efforts by the utility. Inside customers may be required to make up revenue differences.
- Outside customers fail to pay, requiring the utility to either take legal action or write off the account receivable as uncollectible. Inside customers may be required to make up revenue differences.
- Outside customers elect to use another service provider or develop their own utility, requiring the municipal utility to recover abandoned capacity-related operating and capital costs from inside customers.

- Outside customers may challenge the validity of municipal utility rate increases before the state regulatory commission or may sue over policy decisions or perceived discrepancies in service levels or rates, causing the utility to incur legal costs to defend itself.

5.0 FUNDAMENTAL COST-ALLOCATION APPROACHES

The wastewater utility industry uses two basic cost-allocation approaches: the design-basis approach, which allocates costs based on the design of system components, and the functional-basis cost-allocation approach, which allocates costs based on the function each process serves. Chapter 6 discusses each approach in detail. Either approach can be used to allocate costs using the utility basis. The example outlined in Tables A.1 through A.8 at the end of this appendix follow the design-basis cost-allocation approach.

6.0 ALLOCATION OF NET FACILITY INVESTMENT AND PAYMENTS IN LIEU OF TAXES TO COST COMPONENTS

Table A.2 provides an example of depreciated original cost facility investment allocated to cost components that recognizes predominant design criteria by facility as shown in Table 6.6. Cost components consist of volume, capacity, total suspended solids (TSS), biochemical oxygen demand (BOD), total Kjeldahl nitrogen (TKN), customer billing, and customer. As discussed in Chapter 6, the resulting facility allocation in Table A.2 provides an example for next allocating utility-basis revenue requirements of PILOT, depreciation, O&M, and return to cost components.

6.1 Payments in Lieu of Taxes

Payments in lieu of taxes, from Table A.1 and seen at the bottom of Table A.2, are apportioned to cost components based on the total original cost of aboveground facility investment. The allocation of PILOT should follow the basis on which the payment is levied. If based on asset value, as ad valorem taxes are, PILOT should be allocated to cost components in the same manner as aboveground facilities. A separate summary of facility value will need to be developed using only aboveground assets. If the payment is based on gross receipts, then PILOT should be allocated to customers based on class revenue.

TABLE A.2 Allocation of Original Cost Less Depreciation Net Facility Investment and Payments in Lieu of Taxes to Cost Components Using Design-Basis Cost-Allocation Methodology

	Total $	Volume $	Capacity $	Wastewater strength: Total suspended solids $	Wastewater strength: Biochemical oxygen demand $	Wastewater strength: Total Kjeldahl nitrogen $	Customer billing $	Customer $
Facility in service								
Influent pumping	7,558,000		7,558,000					
Preliminary treatment	5,434,000		5,434,000					
Primary sedimentation	3,327,000	3,327,000						
Aeration	11,435,000				11,435,000			
Nutrient removal	2,082,000					2,082,000		
Secondary sedimentation	9,318,000	9,318,000						
Chlorination	1,690,000		1,690,000					
Solids handling and treatment	19,049,000			8,572,000	7,620,000	2,857,000		
Solids dewatering and disposal	28,788,000			12,955,000	11,515,000	4,318,000		
Outfall diversion sewer	1,101,000		1,101,000					
General treatment	3,751,000	528,000	659,000	899,000	1,278,000	387,000		
Total facility	93,533,000	13,173,000	16,442,000	22,426,000	31,848,000	9,644,000	0	0
Pumping and lift stations	4,092,000		4,092,000					
Collection system	79,723,000		71,751,000					7,972,000
General facility	3,986,000	222,000	1,555,000	378,000	537,000	163,000	997,000	134,000
Total facility investment	181,334,000	13,395,000	93,840,000	22,804,000	32,385,000	9,807,000	997,000	8,106,000
Less contributed facilities	49,880,000	0	45,097,000	0	0	0	0	4,783,000
Total net investment	131,454,000	13,395,000	48,743,000	22,804,000	32,385,000	9,807,000	997,000	3,323,000
Payments in lieu of taxes	975,000	103,000	374,000	175,000	248,000	75,000	0	0

6.2 Depreciation

Depreciation, or annual depreciation expense from Table A.1, is allocated to cost components in Table A.3 on the same design basis as the depreciated original cost value for corresponding system assets. Depreciation represents a return of capital investment in system facilities to the owner over time. When fixed asset record details permit, projected test year depreciation associated with the owning utility's facilities should be developed from individual depreciation rates by existing fixed asset category. These rates are applied to projected original gross facility value.

In addition to owner-constructed assets, many utilities own assets that they did not finance or construct, such as assets funded by grants or those contributed by developers or customers. Typically, while investor-owned utilities are prohibited from recovering depreciation on contributed assets, municipally owned utilities have more flexibility. With most contributed property, municipal utilities acquire ownership with the understanding that they will operate, maintain, repair, and replace the assets. Utilities that have received federal grants for water resource recovery facility (WRRF) construction are required to establish user-charge systems that, at minimum, recover operation, maintenance, and replacement capital costs from customers served. In both cases, depreciation can serve as a surrogate for replacement costs incurred by municipally owned utilities for contributed or grant-funded assets.

Municipally owned utilities with outside or wholesale customers must consider carefully treatment of contributed or grant-funded assets. Typically, municipalities exclude depreciation on contributed assets when determining the revenue requirement for outside and wholesale customers, under the logic that depreciation represents the recovery of capital investment cost over time and there is no cost to be recovered for contributed assets. When the assets are replaced using owner-supplied funds, depreciation can be included reasonably in the revenue requirement for outside and wholesale customers. This would also apply to wholesale customer contributions, up-front payments, or other cost-sharing arrangements. Depreciation on assets funded directly by wholesale customers, typically, would be excluded from their revenue requirement to avoid double recovery of these costs.

6.3 Operation and Maintenance Expenses

Operation and maintenance expenses, a significant revenue requirement of a wastewater utility, are allocated to cost components in Table A.4 similarly to facility investment, as shown in Table A.2. The design-basis cost-allocation approach recognizes that equipment and system components, which have to be operated, maintained, repaired, or replaced, are sized to meet design

TABLE A.3 Allocation of Depreciation to Cost Components Using Design-Basis Cost-Allocation Methodology

				Wastewater strength				
	Total $	Volume $	Capacity $	Total suspended solids $	Biochemical oxygen demand $	Total Kjeldahl nitrogen $	Customer billing $	Customer $
Facility in service								
Influent pumping	210,000		210,000					
Preliminary treatment	151,000		151,000					
Primary sedimentation	92,000	92,000						
Aeration	317,000				317,000			
Nutrient removal	58,000					58,000		
Secondary sedimentation	258,000	258,000						
Chlorination	47,000		47,000					
Solids handling and treatment	528,000			238,000	211,000	79,000		
Solids dewatering and disposal	798,000			359,000	319,000	120,000		
Outfall diversion sewer	31,000		31,000					
General treatment	104,000	15,000	18,000	25,000	35,000	11,000		
Total treatment	2,594,000	365,000	457,000	622,000	882,000	268,000	0	0
Pumping and lift stations	113,000		113,000					
Collection system	997,000		997,000					
General	110,000	6,000	43,000	10,000	15,000	4,000	28,000	4,000
Total depreciation	3,814,000	371,000	1,610,000	632,000	897,000	272,000	28,000	4,000

TABLE A.4 Allocation of O&M Expense to Cost Components Using Design-Basis Cost-Allocation Methodology

				Wastewater strength					
	Total $	Volume $	Capacity $	Total suspended solids $	Biochemical oxygen demand $	Total Kjeldahl nitrogen $	Customer billing $	Customer $	Industrial surcharge $
Wastewater treatment									
Influent pumping	636,000	223,000	413,000						
Preliminary treatment	357,000		357,000						
Primary sedimentation	190,000	190,000							
Aeration	968,000				968,000				
Nutrient removal	450,000					450,000			
Secondary sedimentation	319,000	319,000							
Chlorination	393,000	138,000	255,000						
Solids handling and treatment	1,126,000			507,000	450,000	169,000			
Solids dewatering and disposal	2,783,000			1,252,000	1,113,000	418,000			
Laboratory	410,000			185,000	164,000	61,000			
Subtotal treatment	7,632,000	870,000	1,025,000	1,944,000	2,695,000	1,098,000			

Wastewater transmission									
Pumping and lift stations	503,000	176,000	327,000						
Equipment maintenance	770,000		770,000						
Supervision	456,000		456,000						
Subtotal transmission	1,729,000	176,000	1,553,000	0	0	0	0	0	0
Wastewater collection	3,667,000		3,300,000					367,000	
Engineering	795,000	30,000	344,000	114,000	158,000	64,000		53,000	32,000
Industrial waste administration	1,093,000							547,000	546,000
Billing and collection	1,142,000						1,142,000		
All other	640,000	43,000	248,000	82,000	114,000	46,000	46,000	39,000	23,000
Office of the director	502,000	34,000	195,000	64,000	89,000	36,000	36,000	30,000	18,000
Planning and project management	1,002,000	67,000	388,000	128,000	178,000	73,000	71,000	60,000	36,000
Total O&M	18,202,000	1,220,000	7,053,000	2,332,000	3,234,000	1,317,000	1,295,000	1,096,000	655,000
Operation and maintenance reserve	66,000	4,000	26,000	8,000	12,000	5,000	5,000	4,000	2,000
Less revenue from other sources	17,000	1,000	7,000	2,000	3,000	1,000	1,000	1,000	1,000
Net O&M expense	18,251,000	1,223,000	7,072,000	2,338,000	3,243,000	1,321,000	1,299,000	1,099,000	656,000

criteria. Exceptions are purchased power and chemical costs, which should be allocated to the volume cost component to reflect that power consumed throughout the system and designed rates of chemical feed for the treatment processes will be consistently used throughout the year for the total flow handled by the facility. The utility's electric rate schedule may have a demand component, which should be allocated to the capacity cost component.

A more complete discussion about the allocations of specific wastewater utility O&M expense by cost centers to cost components may be found in Chapter 6 related to Table 6.7. Chapter 6 discusses general considerations, the handling of exceptions, and additional considerations during the O&M cost-allocation process. Advantages and approach limitations are also covered.

In Table A.4, net O&M expense is derived by deducting miscellaneous fee operating revenue and non-operating revenue, such as interest earnings, from the allocations of total O&M in proportion to the totals by cost component. Operating fees also may be deducted from specific cost centers before expenses are allocated to cost components.

6.4 Return

Return on facility investment is related to net facility value of the owning utility. Because differential rates of return may be applicable for different classes of customers, particularly between those located inside and outside municipal limits, return should be allocated directly to customer classes, as seen in Table A.5. The applicable rate base to which the municipal utility's rate of return is applied is subject to the same considerations about the nature of service provided to customers as discussed above for depreciation. Namely, like depreciation, municipally owned utilities do not seek to earn a return on contributed assets when establishing rates for outside and wholesale customers. The amount of return allocated to a customer class is dependent on the class's units of service.

7.0 UNITS OF SERVICE

Estimating customer classes' requirements for service or number of units of service by cost component is the next step in the allocation process. Definition of each class and development of class service requirements are discussed in Chapter 7. This estimating process provides two critical allocation elements:

- The total units of service by cost component for the system, which are used to develop unit costs of service, and

TABLE A.5 Customer Class Units of Service Using Design-Basis Cost-Allocation Methodology

Description	Inside retail						Outside retail	System total
	Residential	Commercial	Industrial	Surcharge	Standardized strength	Subtotal		
Wastewater volume (1000 Ccf; 1 Ccf = 2.83 m^3)								
Billed wastewater volume	4473	2761	2197	0	1653	11 084	3617	14 701
Infiltration and inflow allocated two-thirds by customers	2680	282	36		106	3104	1481	4585
Infiltration and inflow allocated one-third by volume	626	387	387		231	1552	507	2059
Total annual treated volume	7779	3430	2541	0	1990	15 740	5605	21 345
Wastewater capacity flowrate (Ccf/d [m^3/d])								
Estimated wastewater peaks	18 382	11 347	9029	0	6793	45 551	14 864	60 415
Infiltration and inflow allocated two-thirds by customers[a]	29 378	3087	391		1159	34 015	16 233	50 248
Infiltration and inflow allocated one-third by volume[a]	6864	4237	3371		2537	17 009	5551	22 560
Total	54 624	18 671	12 791	0	10 489	96 575	36 648	133 223
Wastewater strength (1000 lb, or 454 kg)								
Total suspended solids	7930	4895	3895	5225	6187	28 131	6412	34 543
Biochemical oxygen demand	6811	4204	3346	4321	3609	22 291	5508	27 799
Total Kjeldahl nitrogen	863	533	424	444	412	2676	698	3374
Infiltration and inflow allocated two-thirds by customers[b]	753	79	10	0	30	872	416	1288
Infiltration and inflow allocated one-third by volume[b]	176	109	86	0	65	436	187	623
Customer billing units								
Number of bills	208 400	21 900	8328	828	24 672	264 128	115 150	379 278
Number of customers	52 100	5475	694	69	2056	60 394	28 788	89 182

[a]Infiltration and inflow volumes represent one-third or two-thirds of the volume, from Table 7.4 and a 4.0 peaking factor.

[b]Infiltration and inflow strengths represent one-third or two-thirds of the combined total pounds (kilograms) for total suspended solids, biochemical oxygen demand, and total Kjeldahl nitrogen, shown in Table 7.4.

- Each class's total units of service by cost component that achieve a proportional sharing of all system costs through the uniform application of unit costs of service to each class's unique requirements for service.

Table A.5 shows estimates of retail classes' service requirements inside and outside the municipality's corporate boundaries and in total for the system. Outside retail customers' contributions, shown as one-half of those for inside retail's residential and commercial classes for this example, should reflect the unique requirements of retail and wholesale customers for each utility. Class service requirements consist of directly contributed volumes and strengths plus an estimate of infiltration/inflow (I/I) volume and strengths. Volume requirements have two components: total annual billed amounts contributed and capacity or peak flowrates. Strengths requirements consist of the same strength components that are designed to be handled by the WRRF: TSS, BOD, and TKN. The numbers of customer bills and accounts round out the service requirements.

Typically, WRRF capacities are sized at some multiple of average day flow delivered to the facilities to accommodate both peaks in customer contributed volumes and peaks caused by storm volumes. The difference between annual volumes delivered to the facilities and customers' metered contributions is defined as I/I. Typically, WRRFs are sized up to two times or more their average daily flow, reflecting each facility's unique experience.

For this example, residential, commercial, industrial, and standardized strength classes' capacity requirements are estimated at 1.50 times their respective annual average daily flow. The I/I peak demand is estimated at 4.0 times its average daily flow.

8.0 UNIT COSTS OF SERVICE

Table A.6 shows unit costs of service being developed from the total costs of service by cost component and the system total units of service for the same components. Use of unit cost-of-service provides a defensible position for the utility. All classes of service pay the same unit costs for the same cost component by inside and outside municipal service areas. Differences in allocated costs to customer classes reflect differences in class service requirements and respective unit costs.

In line 1 of Table A.6, total system units from Table A.5 are restated. These are followed in lines 3 to 5 by the cost totals for net O&M expense from Table A.4, depreciation from Table A.3, and PILOT from Table A.2.

TABLE A.6 Unit Costs-of-Service Development Using Design-Basis Cost-Allocation Methodology

Line No.		Total $	Volume $	Capacity $	Wastewater strength: Total suspended solids $	Wastewater strength: Biochemical oxygen demand $	Wastewater strength: Total Kjeldahl nitrogen $	Customer billing $	Customer $	Industrial surcharge $
	Annual units of service									
1	Quantity		21,345	133,223	35,166	28,422	3,997	379,278	89,182	828
2	Units of measure		1,000 Ccf	Ccf/d	1,000 lb	1,000 lb	1,000 lb	bills	each	bills
	Annual operating costs									
3	Net O&M	18,251,000	1,223,000	7,072,000	2,338,000	3,243,000	1,321,000	1,299,000	1,099,000	656,000
4	Depreciation	3,814,000	371,000	1,610,000	632,000	897,000	272,000	28,000	4,000	0
5	Payments in lieu of taxes	975,000	103,000	374,000	175,000	248,000	75,000	0	0	0
6	Subtotals ($)	23,040,000	1,697,000	9,056,000	3,145,000	4,388,000	1,668,000	1,327,000	1,103,000	656,000
7	Unit expense ($/unit)		79.503	67.976	89.433	154.387	417.313	3.499	12.368	792.271
8	Return on facility value	5,796,000								
9	Total facility value ($)	131,454,000	13,395,000	48,743,000	22,804,000	32,385,000	9,807,000	997,000	3,323,000	0
10	Unit value ($/unit)		627.547	365.875	648.467	1139.434	2453.590	2.629	37.261	0
11	Unit inside return ($/unit)		22.137	12.906	22.875	40.194	86.550	0.093	1.314	0
12	Unit outside return ($/unit)		45.497	26.526	47.014	82.609	177.885	0.191	2.701	0.000
	Total unit costs of service									
13	Inside retail ($/unit)		101.640	80.883	112.308	194.581	503.863	3.591	13.682	792.271
14	Outside retail ($/unit)		125.001	94.502	136.447	236.996	595.198	3.689	15.069	792.271

These revenue requirement totals by cost component are divided by total system units for each component to develop unit expenses in line 7 of Table A.6.

The development of return for inside and outside service is handled differently from the other revenue requirements. First, the unit facility values in line 10 of Table A.6 are determined from net facility investment by cost component developed in Table A.2 and the total system units of service from the top of the Table A.6. Then, for inside customers, the 3.53% rate of return from Table A.1 is applied to the unit facility values by cost component to develop the units for inside return in line 11 of Table A.6. Similarly, for outside customers, the 7.25% rate of return from Table A.1 results in the units of outside return in line 12 of Table A.6 when applied to the unit facility values. Total unit costs of service for inside and outside service areas in lines 13 and 14 consist of the unit expenses by cost component plus corresponding inside and outside units of return.

9.0 DISTRIBUTION OF COSTS TO CUSTOMER CLASSES

Revenue requirements to be recovered by general charges and rates for service, originally expressed as O&M, depreciation, PILOT and return, are now in a form that may readily be distributed to customer classes, as shown in Table A.7. Application of the unit costs of service from Table A.6 to the service requirements for each class, or the units of service from Table A.5, results in a dollar amount for each cost component that defines what it costs the utility to provide service for the year to the class by using utility assets related to that cost component. Because the totals of all classes' service requirements are used in the development of unit costs, each class will receive, through this method, a proportional share of the total dollars allocated to each cost component based on their individual service needs. The total of all amounts allocated to each class demonstrates what it costs the utility to serve the class for the year.

Use of a differential in rates of return results in a shifting of risk-related cost-recovery responsibility from inside to outside customers. Because this example depicts municipal utility retail service being provided outside the corporate limits, outside rates should be higher for outside customers than for inside customers. From Tables A.7 and A.5, the average of total allocated dollars per hundred cubic feet (Ccf) of use for outside retail (\$7,797,000 ÷ 3 617 000 Ccf) equates to \$2.16/Ccf (1 Ccf = 2.83 m^3) and is approximately 30.0% higher than the corresponding \$1.69/Ccf for inside retail (\$15,931,000 ÷ 9 431 000 Ccf) (excluding the surcharge and standardized strength customers). The utility-basis cost-allocation process

TABLE A.7 Costs-of-Service Allocations to Customer Class Using Design-Basis Cost-Allocation Methodology

	Total $	Volume $	Capacity $	Wastewater strength: Total suspended solids $	Wastewater strength: Biochemical oxygen demand $	Wastewater strength: Total Kjeldahl nitrogen $	Customer billing $	Customer $	Industrial surcharge $
Total unit costs of service									
Inside retail ($/unit)		101.640	80.883	112.308	194.581	503.863	3.591	13.682	792.271
Outside retail ($/unit)		125.001	94.502	136.447	236.996	595.198	3.689	15.069	792.271
Allocation to classes									
Inside residential									
Units of service		7,779	54,624	8,239	7,121	1,173	208,400	52,100	0
Costs of service ($)	9,571,000	790,000	4,418,000	925,000	1,386,000	591,000	748,000	713,000	0
Inside commercial									
Units of service		3,430	18,671	4,957	4,267	595	21,900	5,475	0
Costs of service ($)	3,700,000	349,000	1,510,000	557,000	830,000	300,000	79,000	75,000	0
Inside industrial									
Units of service		2,541	12,791	3,927	3,378	456	8,328	694	0
Costs of service ($)	2,660,000	258,000	1,035,000	441,000	657,000	230,000	30,000	9,000	0
Inside surcharge									
Units of service		0	0	5,225	4,321	444	828	0	828
Costs of service ($)	2,311,000	0	0	587,000	841,000	224,000	3,000	0	656,000
Standardized strength									
Units of service		1,990	10,489	6,219	3,641	444	24,672	2,056	0
Alloted annual costs ($)	2,797,000	202,000	848,000	698,000	708,000	224,000	89,000	28,000	0
Outside retail									
Units of service		5,605	36,648	6,598	5,691	884	115,150	28,788	0
Costs of service ($)	7,797,000	701,000	3,463,000	900,000	1,349,000	526,000	424,000	434,000	0
Total costs of service ($)	28,836,000	2,300,000	11,274,000	4,108,000	5,771,000	2,095,000	1,373,000	1,259,000	656,000

provides cost-based justification for differences in rates for inside and outside service areas.

In addition, unit costs from Table A.6 may be used in extra-strength surcharge formulas to recover allocated costs from surcharge customers based on their contributed strengths above the limits of domestic strength wastewater.

10.0 COMPARISON OF DIFFERENTIAL AND SYSTEM RATES OF RETURN

Frequently, outside customers contend that there is no risk to the municipal utility to serve outside the municipality's corporate limits, particularly when there is no other wastewater service provider from which these customers may obtain service. These customers may also contend that inside and outside customers should pay the same average costs for service, especially when, as used in this appendix's example, the municipal utility is only extending retail service into adjoining areas using collection system facilities that it already owns. However, there are many examples where political bodies representing outside customers have condemned and purchased WRRFs outside municipal corporate limits, developed their own WRRFs, set up separate utility service for their constituents, and left municipal utilities with unused system capacity costs that inside customers paid for through their rates.

Table A.8 compares the results of using the differential inside and outside rates of return with the results from applying the overall system return, as developed in Table A.1, to all customers.

TABLE A.8 Comparison of Differential and System Returns Using Design-Basis Cost-Allocation Methodology

Customer classes	Inside/outside return differential $	Inside/outside system return $	Difference $
Inside retail			
Residential	9,571,000	9,957,000	(386,000)
Commercial	3,700,000	3,865,000	(165,000)
Industrial	2,660,000	2,782,000	(122,000)
Surcharge	2,311,000	2,393,000	(82,000)
Standardized strength	2,797,000	2,925,000	(128,000)
Outside retail	7,797,000	6,911,000	886,000
Total	28,836,000	28,836,000	0

The system return represents average cost sharing because both inside and outside customers would pay the same unit costs of service. The results from Table A.7 are shown in the inside/outside return differential column. The difference column shows the extent to which inside retail customers will benefit from cost-recovery shifting and the additional amounts each class would have to cover on an average cost basis. The approximately $886,000 difference, compared to the inside/outside return differential column, represents approximately 3.1% of annual revenue requirements, approximately 4.7% of the inside retail classes' total cost allocation, and approximately a 11.4% difference in costs recovered from outside retail.

11.0 SUMMARY

Table A.8 indicates that utility decisions regarding "who should pay for service?" and the methodology used to allocate annual utility costs have potentially wide-ranging effects on all customers. However, as discussed in this appendix, the utility basis provides a sound and court-tested methodology for determining customer class cost recovery responsibility for a variety of retail and wholesale service scenarios where municipal utilities serve customers outside their corporate limits.

Index

www.ingramcontent.com/pod-product-compliance
Lightning Source LLC
Chambersburg PA
CBHW082043130825
31011CB00017B/208
* 9 7 8 1 5 7 2 7 8 4 7 9 6 *